SP Pvt. Ltd.

Natural Resource Management
for Accelerating Agricultural Productivity

by
B. Venkateswarlu
A.K. Singh
Ch. Srinivasa Rao
Gouranga Kar
Ashwani Kumar
S.M. Virmani

2012

Studium Press (India) Pvt. Ltd.

Natural Resource Management
for Accelerating Agricultural Productivity

First published in 2012

ISBN: 978-93-80012-36-0

Citation:

Venkateswarlu, B., Singh, A.K., Srinivasa Rao, Ch., Kar, G., Kumar, A., and Virmani, S.M. (2012) *Natural Resource Management for Accelerating Agricultural Productivity*, **Studium Press (India) Pvt. Ltd**, New Delhi, India. p. 234.

Published by:

Studium Press (India) Pvt. Ltd.
4735/22, 2nd Floor, Prakash Deep Building
(Near Delhi Medical Association),
Ansari Road, Darya Ganj, New Delhi-110 002
Tel.: + 91-11-43240200-15 (15 lines); Fax: 91-11-43240215
E-mail: studiumpress@gmail.com

Printed at:
Salasar Imaging Systems, Lawrence Road, Delhi-35

Prof. M S Swaminathan
Member of Parliament
(Rajya Sabha)

503, Brahmaputra Apartments
Dr. B.D. Marg
New Delhi - 110 001.
Tel: 011 - 2376 6897
E-mail : swami.ms@sansad.nic.in

FOREWORD

I am happy that Dr A.K. Singh and his colleagues have brought out this very timely publication on "Natural Resource Management for Accelerating Agricultural Productivity". Our human population is already over 1.2 billion. In addition we have over 1 billion farm animals which require water, feed and fodder. At the same time, land is becoming a shrinking resource for agriculture, since alternative land uses for industry, housing, roads and business activity are claiming more and more land, much of it belonging to the prime farm land category. In the future, we will have to produce more and more food grains and other farm commodities from diminishing per capita arable land and irrigation water resources. Thus we need to promote a land and water care movement designed to achieve both efficient land and water use. This is the pathway to an ever-green revolution leading to an enhancement in productivity in perpetuity without ecological harm.

The National Agricultural Research System in India has been focusing on natural resource management for more than five decades. Several useful practices and technologies have been developed. Technologies are also available to prevent land degradation, depletion of the aquifer, loss of biodiversity and decline in soil organic carbon content. We should now promote the concept of a "water harvesting pond, a biogas plant and a few fertilizer trees in every farm in the rainfed areas". This will also help to contribute to climate change mitigation and adaption. What we need is now a massive effort in the field of transferring the available technologies to farmers. We should launch a "Bridge the Yield Gap Movement" in order to ensure that the wide gap now existing between potential and actual yields even with the technologies on the shelf, is bridged. I hope this book will be widely read and used by policy makers, scientists, scholars and the general public. We owe a deep debt of gratitude to Dr AK Singh for this labour of love for natural resource conservation and sustainable use.

Date: 11th December 2011
Place: New Delhi

(M S Swaminthan)

PREFACE

Scientific management of natural resources is critical for accelerating agricultural productivity. Although there are several publications dealing with the natural resources and their importance in agriculture, a need has been felt since long to bring out a comprehensive treatise covering all aspects of natural resources like climate, soils, water and bio-diversity; key issues in their management and technologies available for enhancing productivity in crops, cropping and farming systems with an end to end approach. Such a publication will be highly useful if it is based on contemporary scientific research, case studies from the field and critical review of the available knowledge.

In this publication, an attempt has been made to bring out the importance of natural resources for agricultural production, the current status of the key resources like climate, soil, water and bio-diversity, critical issues and concerns in its management and a thorough description of all available technologies from land use planning to soil fertility management, management of problem soils, water harvesting, drought management and contingency planning. New and emerging areas like conservation agriculture, rehabilitation of degraded lands and adaptation to climate change have also been described in detail. One of the chapters, describes all the integrated farming systems modules for different agro-ecosystems like rainfed, irrigated, hill and mountain and coastal. These farming systems can serve as viable alternative strategies to optimize income from unit land by prudent use of natural resources and recycling of the products.

An exhaustive bibliography has been provided for each chapter for further reading. Good quality illustrations and maps are provided for easy understanding of the subject by both the teachers and students and each chapter has been thoroughly reviewed to include all contemporary and authentic information on the particular topic besides providing an overview. We are confident that this publication will be extremely useful to the scientists, teachers, students, NGOs, planners and all other stakeholders who are interested in natural resource management and agriculture production in the country.

B. Venkateswarlu
A.K. Singh
Ch. Srinivasa Rao
Gouranga Kar
Ashwani Kumar
S.M. Virmani

ABOUT THE AUTHORS

Dr. B. Venkateswarlu holds a Doctorate in Soil Microbiology and is currently the Director of Central Research Institute for Dryland Agriculture (CRIDA), Hyderabad, India. He is leading the national research programs of the Indian Council of Agricultural Research (ICAR) on rainfed agriculture and climate change. His research experience is mainly on quantifying the impacts of abiotic stresses of biological nitrogen fixation (BNF) in legumes and understanding plant-microbe interactions in arid and semi arid ecosystems. Dr. Venkateswarlu published over 150 research papers on abiotic stress management and crop nutrition in drylands and authored six books. He has participated in several academic visits to US, UK, Canada, Australia and China and received many national awards for his work on rainfed farming and abiotic stress management. He is a Fellow of the National Academy of Agricultural Sciences of India and serving on key committees set up by the Government in the areas of drought management, rainfed farming and climate change.

Prof. Anil Kumar Singh acquired his M.Sc. (1972) and Ph.D (1977) degree from the prestigious IARI, New Delhi; did his Post Doctoral Research as DAAD Fellow in University of Hohenheim, Germany (1982–83) and IRRI, Manila (1993–95). He obtained a Diploma on *"On Farm Water Management"* from the University of Arizona, Tucson, USA in 1992. *Work Experience*: Scientist S-1, 1976–82, and Scientist S-2/Senior Scientist, 1982–89, Indian Agricultural Research Institute, New Delhi; Principal Scientist & I/charge Soil Physics Unit, Indian Institute of Soil Science, Bhopal, 1989–90; Principal Scientist, 1990–99 and Project Director, Water Technology Centre, IARI, New Delhi,1999 to 2007; Acting Director, IARI, New Delhi 2005–06. Currently, holding the position of Deputy Director General (NRM), ICAR, New Delhi since September 2007. Prof. Singh has been working on increasing crop production and water use efficiency through soil and water management in a career spanning more than three decades. He has contributed significantly in modeling crop responses, nutrient losses through runoff in watersheds as well as developing decision support systems on various aspects of water management. He has worked upon understanding the soil-water-solute interactions, dynamics of nutrients in soil including fertigation systems. He was a member of the team which identified four spectral bands for identifying water stress in wheat, mustard and chickpea for the Indian Remote Sensing (IRS) satellite. He initiated the first systematic studies on aerobic rice cultivation in the country and has demonstrated the potential of aerobic rice production systems in water savings on the farmer's fields'. His contribution to teaching was recognized by IARI Best Teacher Award and Bharat Ratna Dr. C. Subramaniam Best Teacher Award by ICAR. As Deputy Director General (Natural Resource Management) in the Indian Council of Agricultural Research (ICAR), his major responsibilities include developing policy guidelines, setting up research priorities, overall supervision, monitoring and review of all natural resources related programmes in the country as well as developing linkages with other national and international agencies.

Dr. Ch. Srinivasa Rao, Principal Scientist (Soil Science), CRIDA, did his M.Sc (Ag) from Agricultural College, Bapatla (ANGRAU), Ph.D from IARI, New Delhi, Post-doctoral Research at Tel-Aviv University, Israel and was on 3 years deputation at ICRISAT, Patancheru. He joined in Agricultural Research Service in 1992 and worked at various ICAR institutes (IISS, Bhopal, IIPR Kanpur and CRIDA) during last 20 years. In recognition of his outstanding research work, he was elected as fellow of National Academy of Agricultural Sciences (NAAS), Indian Society of Soil Science (ISSS) and Indian Society of Pulse Research and Development (ISPRD). He is the Chief Editor, Indian Journal of Dryland Agriculture (ISDA) and reviewer for Australian Journal Soil Research, Canadian Journal of Soil Science, Applied Clay Research, Archieves of Agronomy, Geoderma, Soil Use and Management besides many national journals. He was recipient of young scientist award from Indian Society of Soil Science, Indian Council of Agricultural Research, National Academy of Agricultural Sciences and Indian Science Congress Association. He is also recipient of Dr.BC Deb Memorial Award and Pran Vohra Memorial Awards from Indian Science Congress Association (ISCA) and ISPRD recognition Award, International Potash Institute-FAI award, PPIC-FAI Award, Dhiru Morajee memorial Awards from FAI and ICRISAT Million Science Award, Doreen Mashler Award from ICRISAT, Sukumar Basu Memorial Award from Indian Agricultural Research Institute FAI Golden Jubile Award and Vasantrao Naik Award from Indian Council of Agricultural Research. He extensively worked on soil potassium dynamics and emerging nutrient deficiencies in dryland soils, soil carbon sequestration, conservation agriculture, participatory soil health management, GHGs in rainfed agriculture and published about 165 research papers in referred journals desides 32 books/bulletins.

Dr. Gouranga Kar, Principal Scientist, soil physics and soil water conservation at Directorate of Water Management, Bhubaneswar obtained his Ph.D. degree from Indian Agricultural Research Institute, New Delhi in the subject of agricultural physics. He joined agricultural research service in 1996 as a scientist, Agro-meteorology and contributed significantly in the field of climate change, hydrology, watershed management, remote sensing, Geographic Information System (GIS), agro-climatic analysis and crop modeling through 29 in-house/sponsored/consultancy projects. He has published 67 research papers in reputed journals and written 23 books/manuals/research bulletins. He has a number of recognitions, national and international, to his credit for his contributions in agricultural science and technology. He was honoured as 'National Fellow' by the Indian Council of Agricultural Research and 'Associate' by the National Academy of Agricultural Science. He was awarded Norman Borlaug international science and technology award and Fulbright Senior Research Fellowship of USA and also recipient of some prestigious recognitions like 'Hari Om Ashram Trust Award' of ICAR, 'Golden Jubilee Commemoration Young Scientist Award' of Indian Society of Soil Science, 'Dr. K.G.Tejwani Award' of Association of Soil and Water Conservationists, 'Vasant Rao Naik Award' of ICAR, 'Sukumar Basu Memorial Award' of IARI, 'Young Scientist Award' of Indian Science Congress Association'.

Dr. Ashwani Kumar, Director, Directorate of Water Management, Bhubaneswar, is an eminent Agricultural Engineer (Soil and Water Conservation Engineering) and outstanding academician with rich experiences in research, teaching, technology dissemination, capacity building, monitoring and evaluation, administration, finance management and intellectual property right (IPAR) related issues. He is presently chairman of Member of Bureau of Indian Standards for Irrigation and Drainage equipment and systems sectional committee, FAD 17, for which he has also worked as member for the last 20 years. He worked as Convenor of working group on Mechanized and Micro

Irrigation, Indian National Committee on Irrigation and Drainage (ICID). He was coveted with several awards such as University Gold medal, Shankar Memorial Award, appreciation certificate of Indian Society of Agriculture Engineer, Best Agricultural Scientist (Rajiv Gandhi State level Sadbhavana Award, 2007), Indira Gandhi Rajbhasha Puraskar Award for the year 2005–06, Prestigious honour, Govind Gupta Memorial Running Award for 2007. Rajbhasha Gaurav Samman by Bharatiya Rajbhasha Parishad, 2007 and ICAR Hari Om Ashram Award for Biennial 2007–08. He is the Fellow of the Institution of Engineers, beside life member to many professional societies related to water management in India.

Dr. S.M.Virmani, born in Bannu, India on January 28, 1938. Educated at Government High School, Ludhiana, Punjab, 1951–54; Agricultural College Ludhiana, Punjab, 1954–61; Indian Agricultural Research Institute, New Delhi 1961–64; National Institute of Agronomy, Paris, France 1964–65; B.Sc. 1958; M.Sc. 1960; Ph.D. 1964; Post-Doctorate 1965. Served as Professor of Soil Science and Chief Scientist (Dryland Agriculture), Haryana Agricultural University, 1971–75; Visiting Scientist, Canadian Department of Agriculture 1973; Visiting Research Fellow, Farming Systems Research Program, 1975–76, Principal Scientist (Agroclimatology) Farming Systems Research Program, 1977–81 and Leader, Farming Systems Research Program, 1981–85 International Crops Research Institute for the Semi-arid Tropics (ICRISAT), Patancheru, A.P., India, Visiting Professor, Evapotranspiration Laboratory, Kansas State University, Manhattan Kansas, USA, 1984–85; Principal Agro-climatologist, Natural Resources Management Program, 1986–91, and Team Leader, Agronomy, 1991–98 ICRISAT; Coordinator, ADB-assisted Asian Watersheds Research Program 1988–2001. Received Lifetime service achievement award, Soil and Water Conservation Society, India; ICRISAT Dedicated Research Service Awards, 1986 and 1996; Chairman, Commission VI, International Union of Soil Sciences, 1994–1998; Member, American Society of Agronomy; National Geographical Society, USA. Conferred fellow of Indian Society of Soil Science and National Academy of Sciences, India.

CONTENTS

1

INTRODUCTION

Though the contribution of agriculture to GDP has been steadily declining, nearly 60% of the population still directly or indirectly depends on agriculture in India. Indian agriculture has made rapid strides since Independence in achieving a record growth in food production from 51 m tones in 1950–51 to 230 m tones in 2010–11, but the sector continues to face major challenges in sustaining this growth and meet the demands of growing population; maintaining the interest of the farmers in farming in the midst of increasing cost of cultivation, problems of labour availability and uncertain prices. The growth achieved during 70s and 80s was largely driven by high yielding varieties, irrigation and use of chemical inputs. According to Abrol and Sanger (2006), the major R & D efforts in the 'green revolution' era focused on enhancing productivity of selected food grains and some other crops by adopting the above strategies, but the new challenges demand that efficient resource conservation and management should receive high priority to ensure that the earlier grains are sustained and further enhanced. The recent slow down in productivity growth has been attributed to resource degradation problems and falling factor productivity in major production systems.

Climate change has emerged as yet another challenge to Indian agriculture. Direct impacts of temperature increase, changed rainfall pattern and extreme weather events on crops, livestock and fisheries and indirect affects on water availability are threatening our food security. Land and water availability to agriculture are declining in view of the competing demands from other sectors. Shortage of fodder resources and fresh water pollution are affecting the livestock and fishery sectors. Rainfed agriculture which covers 58% of the net sown area continues to face major challenges like droughts, land degradation and poor investment capacity of the farmers. The issue of prudent management of natural resources like land, water and bio-diversity, therefore has come to the centre stage of agricultural research and development in the country. *Kharif* 2009 experienced severe drought in many parts of the country but towards the end of the season heavy rains damaged crops particularly rice due to water inundation. Thus food production in India reduced from 231 million tones in 2008 to 218 million tones in 2009. Besides, low factor productivity in Indian agriculture raised serious concerns in terms of economy as well as environmental point of view. For unit production of food, more and more inputs are needed now. During the year 2010 several important issues related to agriculture have come into focus. Among others they include effect of global warming on agriculture, the depletion of soil organic matter (SOM) and consequent declining response to fertilizers along with the show up of secondary and micro nutrient deficiencies, ecological farming as an alternative to the green revolution, focus on rainfed agriculture, lack of attention on smallholders and the growing anxiety on the seed sector. While there is a greater emphasis on genotypes (crops, livestock's) to perform better inspite of climate change, less attention seems to be given on its effects on the Natural Resources. With the skewed and intense rainfall and concurrent rise in temperatures, the fertile top soil is melting away at a much faster rate. Presently ISRIC, based on time series data from 1983 to 2006, tells us that India is losing 4.9 ha/minute through degradation. This loss might be more in the years to come. The conservation tillage or zero-till may partly mitigate the problem. Improving the soil organic matter could be the other solution. Thus management of natural resources requires a comprehensive and multi pronged approach, involving technology and policy inputs.

2

NATURAL RESOURCES AND AGRICULTURAL PRODUCTIVITY

2.1 CLIMATE

India is a vast country with a total geographical area of 329 m ha and a net sown area of 141 m ha. The key natural resources are: land, biomass and water bodies, both inland and coastal. The country has diverse climates, soils and biodiversity. The climatic and agro ecological classification of the country has been evolving over time.

Initially, Planning Commission has divided the country into 15 Agro climatic zones, whereas as under the National Agricultural Research Project (NARP), 127 agro ecological zones have been identified (Fig. 2.1). More recently, NBSSLUP has divided the country into 20 homogenous agro ecological regions and 60 sub-regions (Fig. 2.2) based on topography, climate, soils and effective growing seasons (Sehagal *et al.*, 1990). In view of the variability of the crop growing conditions under semi-arid and sub-humid conditions, the climatic zones have been regrouped by CRIDA in relation to rainfall, LGP and AESR (Table 2.1). This classification characterizes the suitability and potential of each sub region for a given land use, cropping/ farming system.

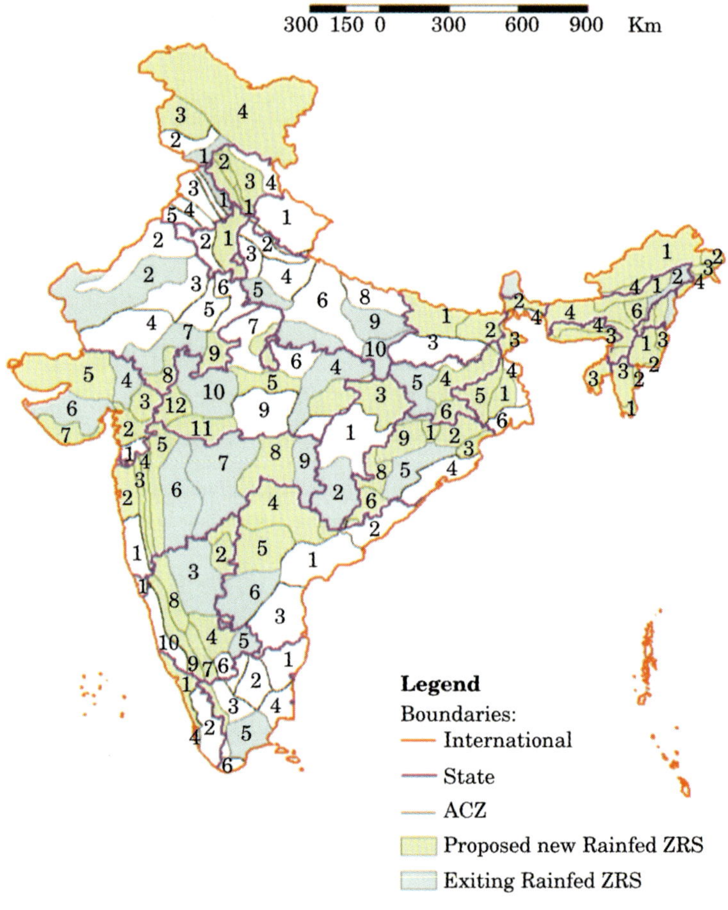

Fig. 2.1: Delineation of NARP Zones

2.2 SOILS

India has diverse soil types falling into eight orders. Taxonomically, Indian soils fall under Entisols (80.1 m ha), Inceptisols (95.8 m ha), Vertisols (26.3 m ha), Aridisols (14.6 m ha), Mollisols (8.0 m ha), Ultisols (0.8 m ha), Alfisols (79.7 m ha), Oxisols (0.3 m ha) and non-classified soil (23.1 m ha). Rainfall wise, 15 m ha area falls in a rainfall zone of <500 mm, 15 m ha under 500 to 750 mm, 42 m ha under 750 to 1150 mm and 25 m ha under >1150 mm rainfall. From practical classification, alluvial soils are the most dominant (93.1 m ha) followed by red soils

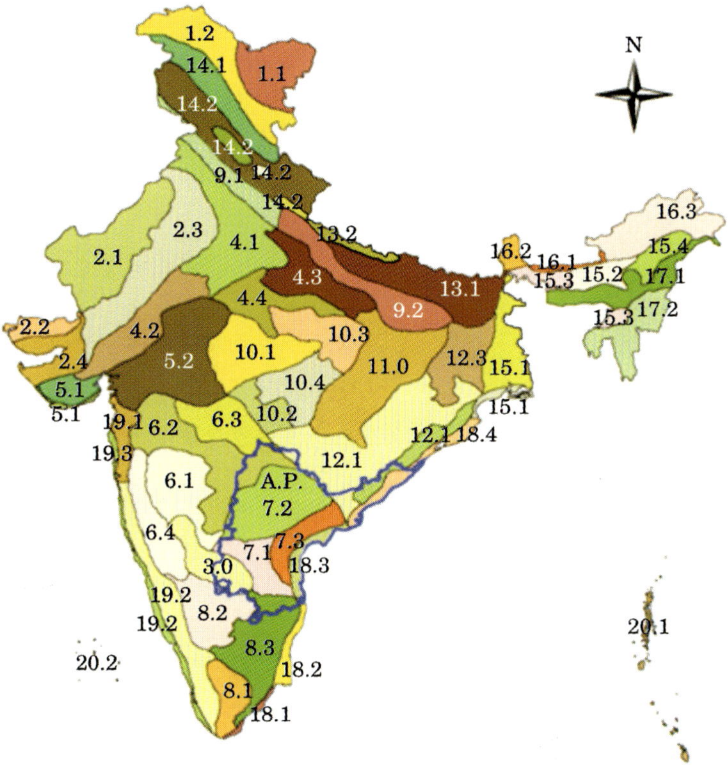

Fig. 2.2: Agro-ecological sub-regions of India (*Source*: NBSS&LUP, Nagpur)

(79.7 m ha), black soils (55.1 m ha), desert soils (26.2 m ha) and lateritic soils (17.9 m ha) (Table 2.2). Both the ecological (supporting biomass, forests, buffering action) and the economic (food, fodder and fibre production) functions of the soils are important in India as the crop and livestock production systems are highly integrated in a farming systems mode.

2.3 WATER

The major source of water to Indian agriculture is the monsoon system. India annually receives about 400 m ha m of rainwater, which is 4% of global precipitation. Out of this, 24 m ha m equivalent rainwater is available for harvesting. Most part of the country receives rains during south-west monsoon except parts of Andhra Pradesh and Tamilnadu

Table 2.1: Regrouping of agro-ecological sub-regions of the country based on climate types

Type of climate	*Annual rainfall (mm)*	*Moisture index 1M (%)*	*Growing period (days)*	*Agro-ecoregion/ sub-region*	*Physiography*	*Area (m ha)*
Arid						
Cold arid	<500	> –66.7	60–90	1.1, 1.2	Western Himalayas, parts of J & K	14.31
Hot arid	<500	> –66.7	< 90	2.1, 2.2, 2.3, 2.4, 3.0	Western Plains and Kutch Peninsula, Deccan Plateau	38.08
Semi Arid						
Dry	500–700	–66.7 to –55.8	90–120	4.1, 4.2, 5.1, 6.1, 7.1, 8.1, 14.1, 18.1	Northern plains, Central highlands including Aravalis. Deccan plateau, T.N uplands, South Tamilnadu plains	41.57
Moist	750–1000	–55.7 to –33.3	90–150	4.3, 4.4, 5.2, 5.3, 6.2, 6.3, 7.2, 8.2, 8.3, 9.1, 18.2	Indo-Gangetic plains, Bundelkhand up-lands, Malwa plateau, eastern Gujarat plain, Vindhyan hills, Central & Western Maharashtra, North Karnataka, Vidarbha, N. Telangana, Central Karnataka, Tamilnadu plains, Punjab & Rohilkhand	72.15
Sub-humid						
Dry	1000–1200	–33.3 to 0.0	150–180	6.4, 7.3, 9.2, 10.1, 10.2, 10.3, 11.0, 12.3, 14.2, 18.3	Western Karnataka plateau, Eastern Ghats (south), South Bihar plains, Rohilkhand, Malwa plateau, Vindhyan Scariland, Narmada valley, Bundlekhand plateau, Chattisgarh, Mahanadi-basin, Chotanagar plateau, Gujarat hills, South Kashmir and Andhra plain	57.99
Moist	1200–1600	0.0 to 20.0	150–210	10.4, 12.1, 12.2, 13.1, 13.2, 18.4	Satpura range & Wainganga valley, Gujarat hills, Eastern Ghats, North Bihar, Fort hills of Central Himalayas, Utkal plains & East Godavari Delta	35.85
Humid and Per-humid						
Humid	1600–2000	20.0 to 1000.0	180–270	15.2, 18.5, 19.1, 19.2, 19.3, 20.2	Parts of western and eastern Himalayas, Bengal basin and Assam Plains	33.3
Per-humid	> 2000	> 100	> 270	14.3, 14.4, 14.5, 15.3, 15.4, 16.1, 16.2, 16.3, 17.1, 17.2, 20.1	Parts of eastern Himalayas, north eastern hills, western Ghats and coastal plains.	31.3

Compiled by CRIDA (AICRPAM) using NBSS&LUP data

which also receive from the North-East monsoon. Extensive efforts were made in the country to develop water resources for agriculture before independence through construction of village tanks and small harvesting structures. Post independence, the Government of India has developed huge infrastructure of canal based irrigation system. Since 1970s, with the wide availability of electricity and pumping technologies, there was a massive increase in the ground water utilization through tubewells under private investment. Groundwater today provides about 70% of irrigation and 80% of drinking water supplies. The total irrigated area in the country is 63.20 m ha out of which the canal system, tanks and the tube wells + open wells constitute 26.24%, 3.13% and 61.02% respectively (2008–09). During the last 60 years after independence, the management of water resources has changed from the hands of communities (tanks and small water structures) to Government (major and medium irrigation projects), and into the private domain (groundwater). Though, the net irrigated area in the country has reached 60 m ha, there is a large spatial variability in availability of water to agriculture in view of the rainfall pattern, land forms, and the river systems.

Table 2.2: Broad soil groups and their moisture storage capacities of India

Broad soil group	*Sub-group (based on soil depth)*	*Moisture storage capacity (mm)*
Vertisols and related soils	Shallow to medium (upto 45 cm) Medium to deep (45–90 cm) Deep (>90 cm)	135–145/45 cm 145–270/90 cm 300/m
Alfisols and related soils	Shallow to medium (upto 45 cm) Deep (>90 cm)	40–70/45 cm (sandy loam) 70–100/45 cm (loam) 180–200/90 cm
Aridisols	Medium to deep (upto 90 cm)	80–90/90 cm
Inceptisols	Deep	90–100/m (loamy sand) 110–140/m (sandy loam) 140–180/m (sandy loam)
Entisols	Deep	110–140/m (sandy loam) 140–180/m (loam)

2.4 BIODIVERSITY–TREES/PASTURES

Fodder and grazing lands are another important source supporting livestock production in the country. The current fodder production in the country is estimated at 385 and 428 m tons of green and dry fodder respectively, which represents 61% and 22% of the requirement (NRCAF, 2007). Historically, common property resources have served as major sources of grazing lands. With increasing pressure on lands, permanent pastures and grazing lands have declined from 13.26 to 10.45 m ha between 1970-71 and 2003-04. Likewise miscellaneous tree crops and groves, which also support livestock as top feed source, also declined by 22% during the same period. India has total inland water resources of 73.59 lakh ha water spread area for inland fishery production and a long coast line of 7600 km that supports marine fisheries (GoI, 2008). The above account indicates the huge diversity and extent of natural resources base in the country for agriculture, livestock and fishery production.

3

NATURAL RESOURCES MANAGEMENT: KEY ISSUES AND CONCERNS

India has 18% of world's population, 15% of the world livestock with only 2.3% of the geographical area, 4.2% of fresh water resources, 1% forests and 0.5% pasture land. These figures indicate the high pressure on our natural resources. The net sown area in the country has remained constant for many years at 141 m ha, but the human and livestock population has been steadily increasing. Land degradation and declining water availability and deteriorating quality are key concerns. There is also growing competition for land from non-agricultural sectors as evidenced by the whopping 261% increase in land under non-agricultural uses since independence. The resistance of farmers across the country for acquiring land under special economic zones (SEZs) is a clear indication of the growing conflicts and challenges land use poses in future.

3.1 LAND DEGRADATION

Land degradation could be due to physical causes like mining, disposal of industrial wastes and water logging; chemical causes like salinity, alkalinity and soil acidity and natural causes like water and wind erosion. There have been various estimates of degraded lands/waste lands made by different agencies. Recently, all these estimates have been harmonized by adopting spatial data integration in GIS environment (Maji, 2007). The total degraded area is now put at 120.72

m ha out of which 104.19 m ha is arable land 16.53 m ha is open forest land. Out of these, 73.27 m ha is affected by water erosion, 12.40 ha by wind erosion, 5.44 ha affected by salinity and 5.09 ha by soil acidity. Some areas are affected by more than one cause.

Globally, erosion due to water is the single most cause for degradation which is true in India also. An analysis by CSWCRTI and NBSS&LUP revealed that 39% area in the country is having erosion rates of more than permissible limit of 10 t/ha/year. About 11% area falls in very severe category with erosion rates of more than 40 tons/ha/year.

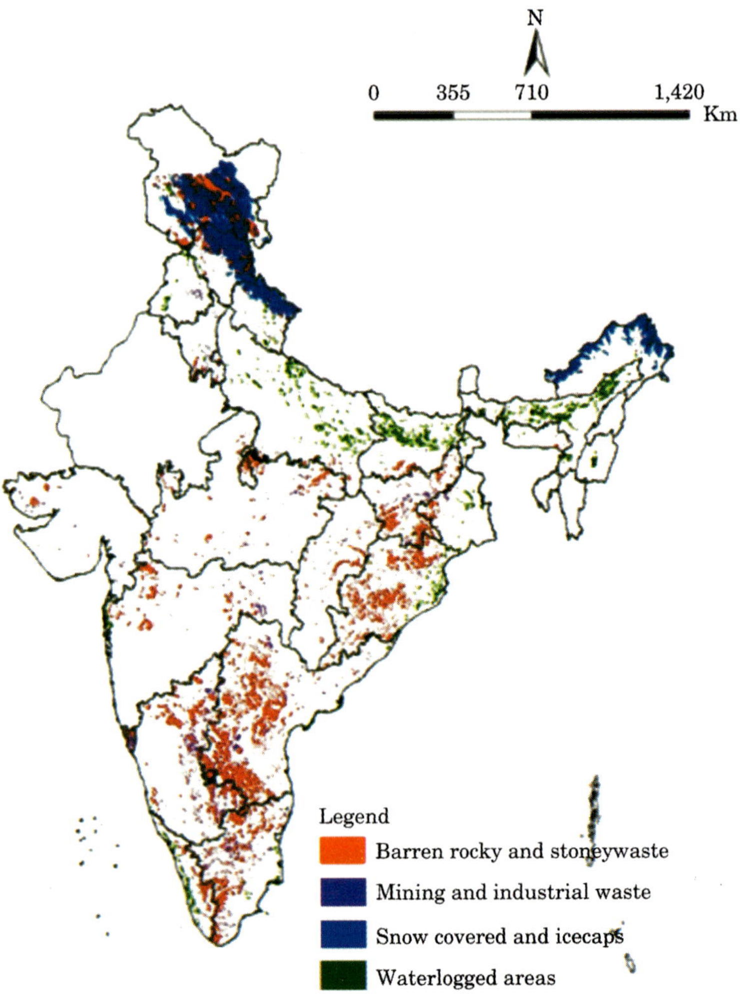

Fig. 3.1a: Physical land degradation in India (*Source*: NRSA, 2005)

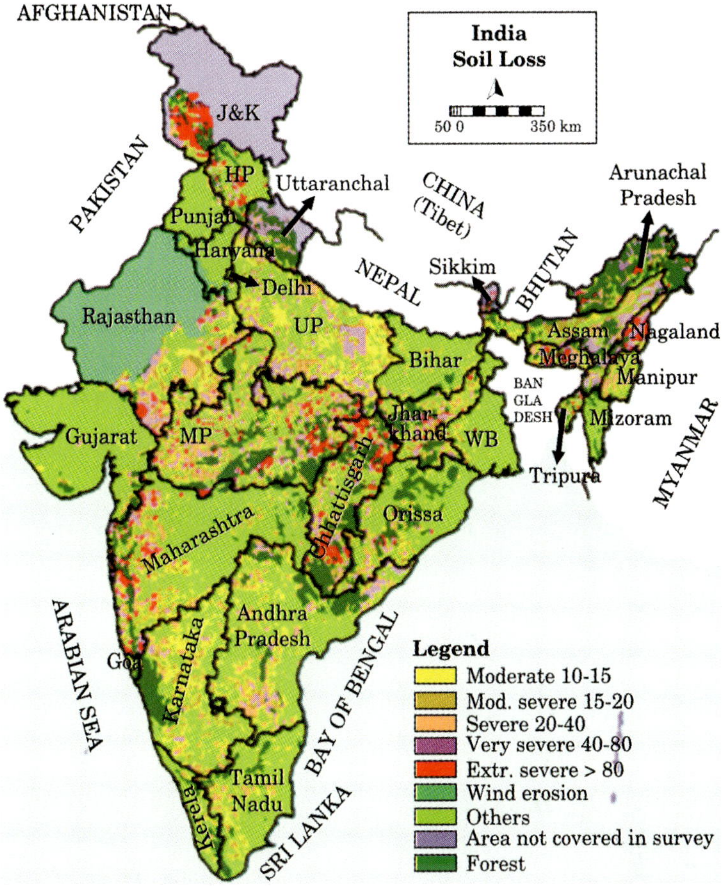

GIS section, NBSS & LUP, Nagpur

Fig. 3.1b: Soil loss by water erosion (>10 tonnes/ha/yr) (*Source*: Maji *et al.*, 2008)

The states in north-west and north-east Himalayas are the worst affected. Erosion rates are also linked to loss of crop productivity which may range between 5–50% in different soil types (Sehgal and Abrol, 1994).

The harmonization process for producing realistic estimates on the degraded and waterlands was thoroughly discussed by the National Remote Sensing Agency, Natural Resource Management (NRM) institutes of the ICAR, scientists of the NAAS and other stakeholders in three, 2-day workshops, held at the National Remote Sensing Agency. Similarities and dissimilarities in the various maps and datasets produced by various organizations were analyzed, and shortcomings

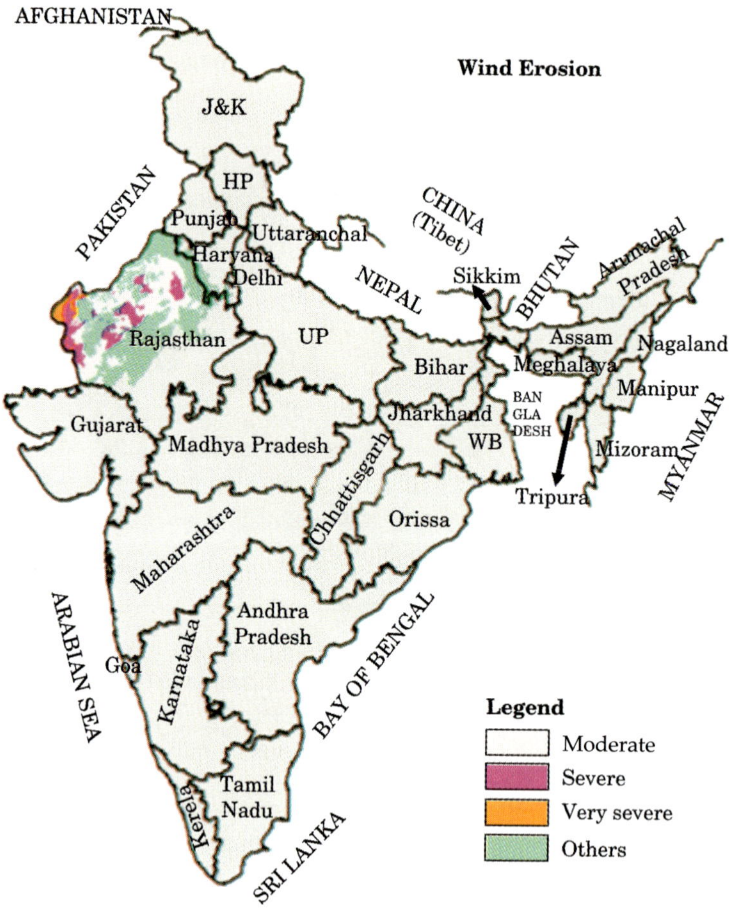

Fig. 3.1c: Wind erosion in India (>10 tonnes/ha/yr) (*Sources*: CAZRI & NBSS & LUP, 2008)

in the status of the degraded and wastelands were pinpointed. The major difference was identified as the scale of maps. Some of the items of the duplicated area, data sources and definitions were harmonized. Some degradation classes were not responsive to amendments or management. For example, theoretically any soil having pH of less than 7 is acidic, but responses to amendments and reclamation were observed in the soil having pH less than 5.5. Similarly snow-covered land or glaciers were included in wastelands in some of the maps. Ice-capped land may not have green vegetation but glaciers are precious; and this important source of water should not be treated as wastelands.

In fact, there is a global concern for preventing excessive melting of glaciers for sustaining water supplies. Rocky areas by definition are not soils and are treated as degraded lands. Such lands are also a source of water for watershed development for supporting biological activities downstream. Forest land with more than 40% canopy has not been considered degraded during harmonization process. Soil erosion below 10 tonnes per hectare generally does not significantly affect productivity and has not been counted as degraded wastelands. The harmonized estimates, therefore, were derived from the point of view of the practicalities of reclamation, amelioration and management for agricultural planning rather than purely as an academic exercise. This has been necessary for a realistic planning of the land reclamation, wasteland management, land-use planning and watershed management for sustainable agriculture.

The harmonization process was made possible by putting all datasets in the GIS-compatible format. In the GIS exercise, the spatial layers on water erosion (soil loss), salt-affected soils, acid soils, lands affected by wind erosion, dense forested and open forest layers were considered. In the first step, the soil loss classes of >10 tonnes/ha/yr (10–15, 15–20, 20–40, 40–80 and >80 t/ha/yr) were considered. These were then grouped, and layer 1 was thereby generated. In this, soils loss classes of very slight (<5 tonnes/ha/yr) and slight (5–10 tonnes/ha/yr) were not considered. As a second step, acid soil layer (<5.5 pH) was superimposed on the soil loss layer and thus layer 2 was generated. In the third step, the wind erosion layer (moderate to extremely severe classes) was superimposed on layer 2 and the layer 3 was generated. Since wind erosion areas do not fall either under dense and open forest cover, the total area remained same. As the fourth step, the dense forest (>40% canopy) and open forest (<40% canopy) (FSI, 1999) was superimposed on the layer 3, and areas of water erosion and acid soils coming under dense forest (>40% canopy) were deducted, and the land degradation classes associated with open forest were generated as a separate class and layer 4 was generated. As the fifth step, the salt-affected layer having two classes (saline and sodic) was integrated with land degradation classes and all possible combination classes were identified to generate layer 5. As the next step, the physical degradation layers like barren rock and stony wastelands, mining and industrial wastelands, snow covered and ice caps and waterlogged areas assessed by the NRSA were integrated with other land degradation layers and the final land degradation map of India with all possible combined classes was generated. These combined classes denote areas with complex problems (more than one problem occurring in the same parcel of land). It should be noted that the harmonization procedure adopted above has a built-in standard deviation of ± 2%. The methodology described above is presented in the form of a flow chart followed for the harmonization procedure (Fig. 3.2).

A harmonization exercise as described before has been done by developing a methodology involving various stakeholder organizations

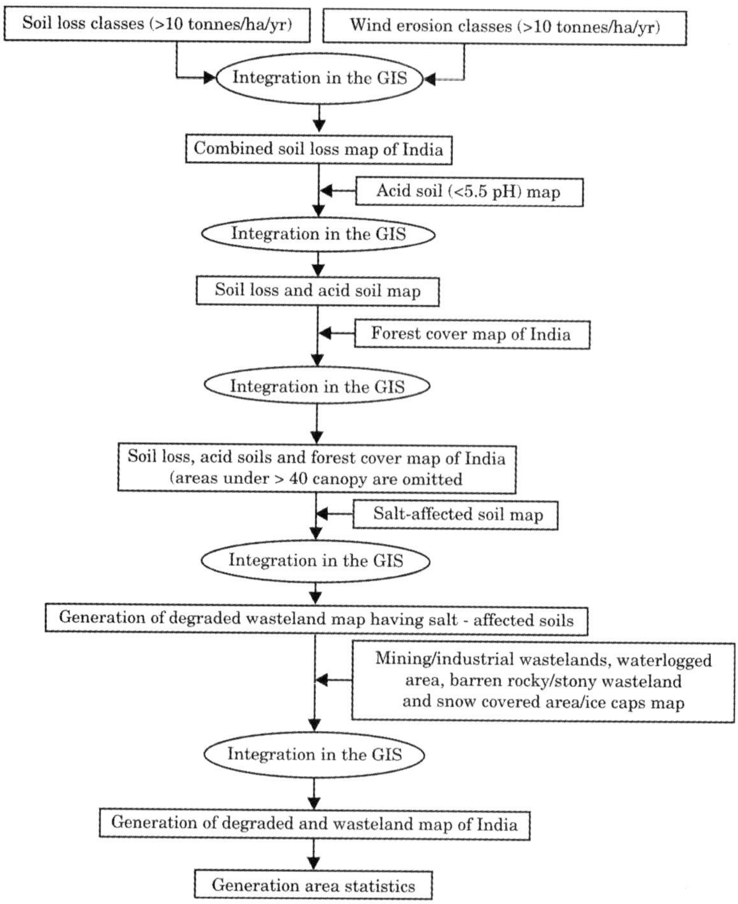

Fig. 3.2: Harmonization process (*Source*: NAAS, 2010)

and a final harmonized degraded and wastelands map of India was generated with all possible combination classes (Fig. 3.3). Degraded

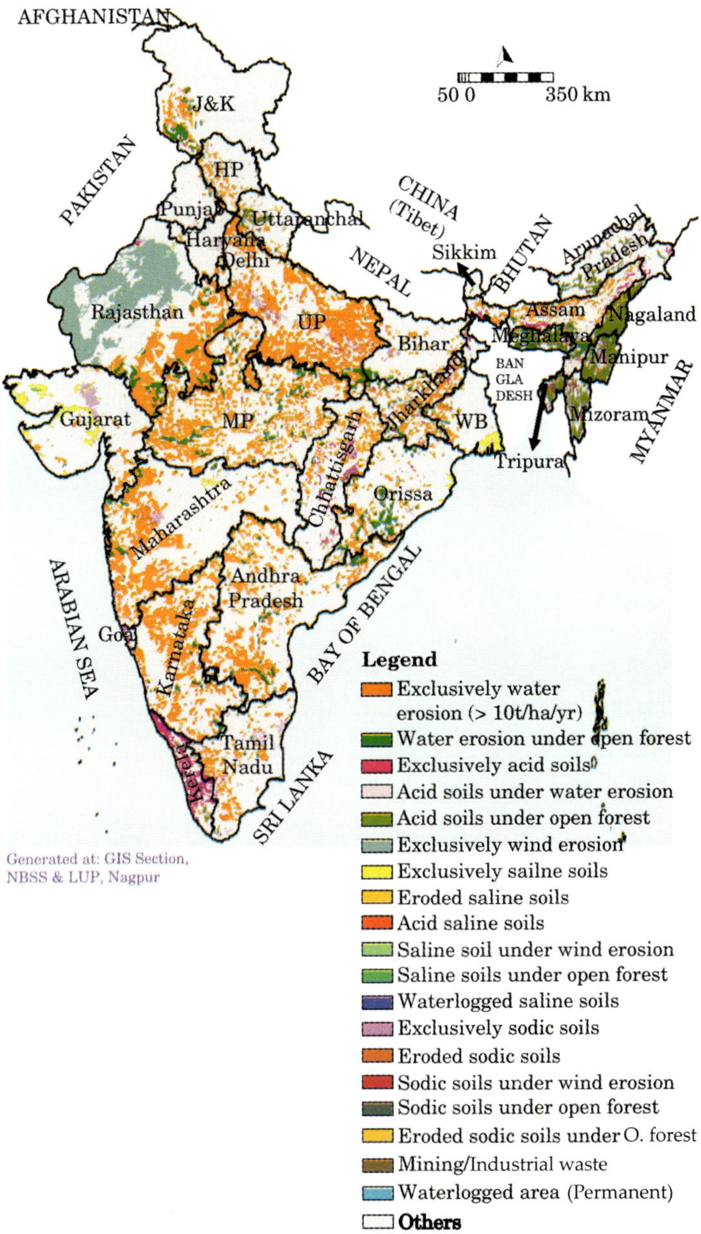

Fig. 3.3: Degraded and wastelands of India (*Source*: NBSS & LUP)

and wasteland areas under water, wind, physical and chemical degradations have been worked out (Table 3.1). Nineteen degradation classes have been identified and described, and area statistics under each class has been assessed (Table 3.1).

Table 3.1: Harmonized area statistics of degraded and wastelands of India (M ha)

Degradation type	*Arable land (Mha)*	*Open forest (<40% canopy) (Mha)*	*Data source*
Water erosion (>10 tonnes/ha/yr)	73.27	9.30	Soil loss map of India CSWCRT & TI
Wind erosion (Aeolian)	12.40	–	Wind erosion map, CAZRI
Sub-total	**85.67**	**9.30**	
Chemical degradation			
Exclusively salt affected soils	5.44	–	National salt affected soils map,
Salt affected and Water eroded soils	1.20	0.10	CSSRI, NBSS & LUP, NRSA and others
Exclusively acidic soils (pH < 5.5)[#]	5.09	–	Acid soil map of India,
Acidic (pH < 5.5) and water eroded soils[#]	5.72	7.13	NBSS & LUP
Sub-total	**17.45**	**7.13**	
Physical degradation			
Mining and industrial waste	0.19		Wasteland Map of the NRSA
Waterlogging (permanent surface inundation)[$]	0.88		
Sub-total	**1.07**		
Total	**104.19**	**16.53**	
Grand Total (Arable land and open forest)	**120.72**		

Notes: Forest Survey of India Map (1999) was used to exclude degraded land under dense forest. (•) *Unculturable Wastelands*: Barren rocky/stony waste: 6 M ha, are source for runoff water and building material, Snow covered/Ice-caps: 6 M ha, are best source of water and are not treated as wastelands. (#) For acid soils, areas under paddy growing and plantation crops were also included in the total acid soils. ($) Sub-surface waterlogging was not considered. *Source:* NBSS & LUP (2009)

3.2 SOIL FERTILITY ISSUES

Declining soil fertility and nutrient imbalance is another major issue affecting agricultural productivity. Organic matter levels have declined sharply in intensively cropped Indo-Gangetic plains leading to stagnant yields of rice-wheat cropping system. In addition to universal deficiency of nitrogen, deficiencies of potassium, sulphur and micro nutrients are increasing. Zinc deficiency has become most acute followed by sulphur and boron (Motsara, 2002). It is estimated that 29.4 m ha of soils in India are experiencing decline in fertility with a net negative balance of 8–10 m t of nutrients per annum (Lal, 2004). Poor nutrient use efficiency is another cause of concern. Indian soil are low in organic matter thus soils are showing 63 per cent low, 26 per cent medium and 11 per cent high status. About 80 per cent soils tested were low to medium in available P. Potassium status was low to medium in 50 per cent soil samples (Table 3.2). Similarly among micronutrients, Zn is most deficient followed by Fe, Mn and Cu (Table 3.3). So far soil fertility issues were mainly focused in irrigated agriculture, but recent studies indicated that drylands are not only thrusty but also hungry (Srinivasarao *et al.*, 2006, 2007, 2008, 2009c; Sharma *et al.* 2008). Most of

Table 3.2: Fertility status of N. P and K in some states of India

State	*No. of samples analyzed*	*Nitrogen*			*Phosphorus*			*Potassium*		
		L	*M*	*H*	*L*	*M*	*H*	*L*	*M*	*H*
Madhya Pradesh	1,38,553	40	41	19	39	38	23	10	32	58
Uttar Pradesh	8,07,424	80	15	5	71	26	3	12	55	33
Maharashtra	93,142	67	26	7	86	12	2	8	18	74
Andhra Pradesh	3,12,521	62	21	17	57	29	14	9	30	61
Karnataka	3,17,213	29	37	34	31	48	21	7	32	61
Orissa	2,51,196	60	23	17	59	28	13	33	41	26
Tamil Nadu	4,91,657	75	16	9	24	41	35	12	36	52
India	36,50,004	63	26	11	42	38	20	13	37	50

Source: Motsara (2002)

Table 3.3: Micronutrient deficiencies in some states of India

State	*No. of samples (range of samples analyzed for different states)*	*Percentage deficiency of available micronutrients*			
		Zn	*Fe*	*Cu*	*Mn*
AP	5219–6563	51	2	1	2
Bihar	17802–19078	54	6	3	2
Gujarat	29532	24	8	4	4
MP	11204–12000	63	3	1	3
Tamil Nadu	19559–20580	53	15	3	8
UP	24425–25122	45	6	1	3
Karnataka	24411–25542	78	39	5	19

Source: Takkar (2006)

the rainfed regions are low in organic carbon and available N and soils (Fig. 3.4) are multi-nutrient deficient (Table 3.4).

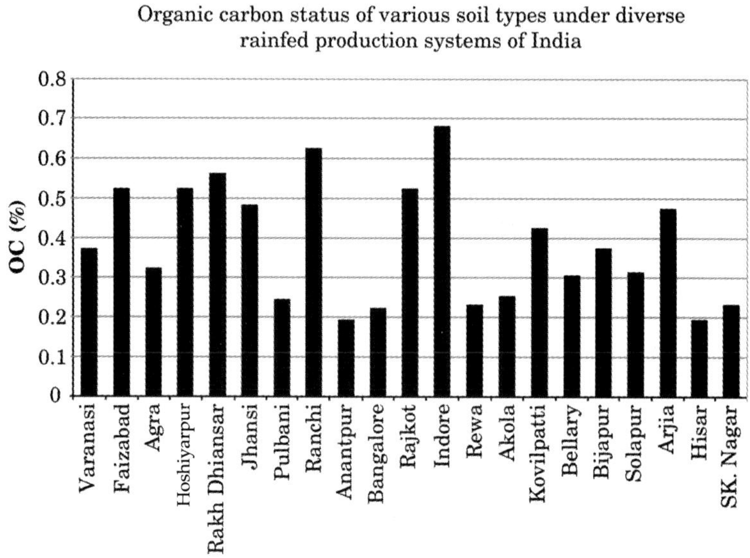

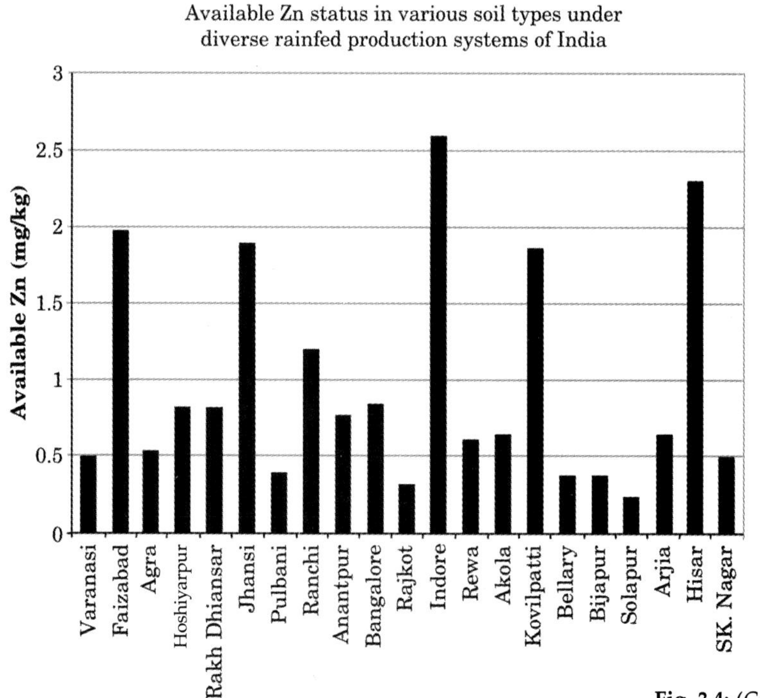

Fig. 3.4: (*Contd...*)

Fig. 3.4: *(Contd...)*

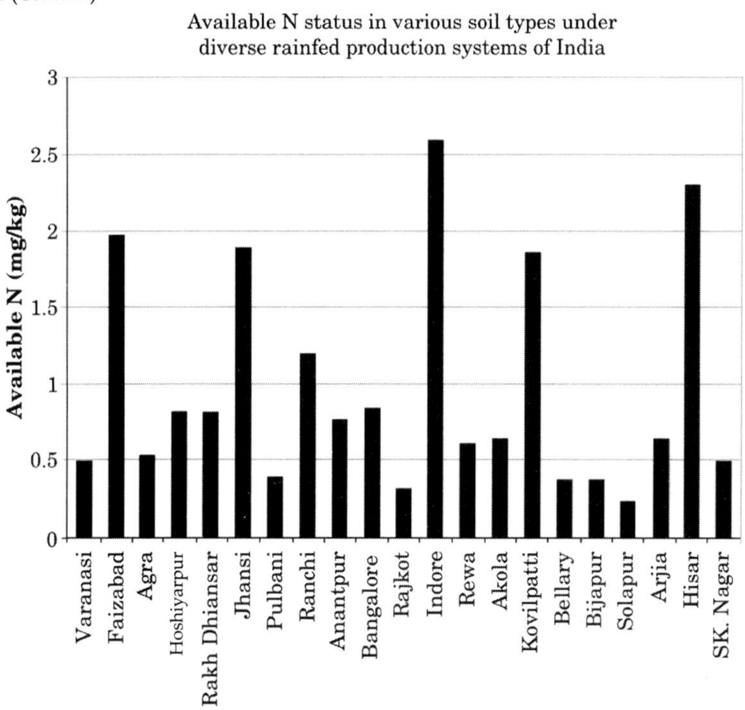

Fig. 3.4: Organic carbon, available N and Zn status of 21 locations of AICRPDA centres. *Source*: Srinivasarao and Vittal (2007).

Table 3.4: Emerging nutrient deficiencies in dryland soils (0–15 cm) under diverse rainfed production system of India

Location	*Limiting nutrient (Low/Deficient)*
Varanasi	N, Zn, B
Faizabad	N
Phulbani	N, Ca, Mg, Zn, B
Ranchi	Mg, B
Rajkot	N, P, S, Zn, Fe, B
Anantapur	N, K, Mg, Zn, B
Indore	–
Rewa	N, Zn
Akola	N, P, S, Zn, B
Kovilpatti	N, P
Bellari	N, P, Zn, Fe
Bijapur	N, Zn, Fe
Jhansi	N
Solapur	N, P, Zn

Table 3.4: *(Contd...)*

Table 3.4: *(Contd...)*

Location	*Limiting nutrient (Low/Deficient)*
Agra	N, K, Mg, Zn, B
Hisar	N, Mg, B
SK. Nagar	N, K, S, Ca, Mg, Zn, B
Bangalore	N, K, Ca, Mg, Zn, B
Arjia	N, Mg, Zn, B
Ballowal - Saunkri	N, K, S, Mg, Zn
Rakh-Dhiansar	N, K, Ca, Mg, Zn, B

Source: Srinivasarao and Vittal (2007)

3.3 SALINITY AND WATER LOGGING

Soil salinity is another constraint limiting productivity on large areas. Salt-affected soils occur extensively in different agro-ecological zones, particularly the arid, semi-arid and the sub-humid regions. The most affected states are Uttar Pradesh, Gujarat, Rajasthan, West Bengal and Andhra Pradesh. The soils are categorized as alkali or saline soils depending upon the nature of soluble salts. The development of the major and medium irrigation projects has led to the occurrence of water logging and secondary salinity in irrigation commands. The area under salinity has been increasing at the rate of 3,000 to 4,000 ha per annum. As per an estimate (Anonymous, 2004), about 5.6 M ha of the salt-affected soils out of the total of 8.5 m ha, are in the irrigation commands. Water-logging of fertile alluviums in eastern India (3.28 m ha) and acidity in about 25 m ha of arable lands in east and north east are other critical constraints affecting productivity (Fig. 3.5).

3.4 WATER MANAGEMENT ISSUES

Water management is another critical area for sustaining agriculture productivity. It is estimated that by 2050 about 22% of geographical area and 17% of the population would be under absolute water scarcity (<500 M^3 per capita per year), and about 70% of the area and 16% of the population will be on the verge of economic scarcity and health risk with water availability of less than 1000 M^3 per capita per year (Planning Commission, 2002). Declining water availability to agriculture and use of poor quality waters are the key constraints. The present water management practices are beset with many problems of wastage and

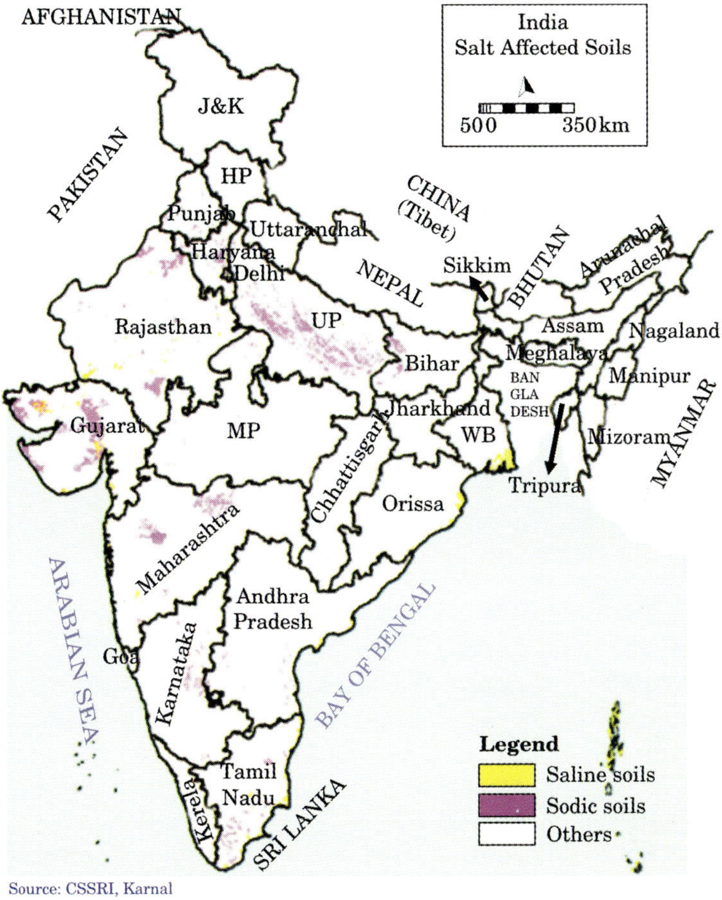

Fig. 3.5: Salt-affected soils of India (*Source*: CSSRI, Karnal)

poor productivity for unit water used. There has been an alarming increase in the exploitation of ground water. Irrigation consumes 92% of the ground water draft. While in Northern, Western and southern states, the ground water is being fully exploited, the Eastern and North Eastern parts offer some untapped potential.

Water quality due to geological factors and pollution are other concerns for irrigated agriculture. High levels of fluoride occurs in about 200 districts and arsenic contamination of ground water in West Bengal and parts of Bihar have become major concerns not only for crop production, but the entire food chain. Nitrate pollution in irrigated regions of Punjab, Haryana and Andhra Pradesh, inland salinity arid regions of Rajasthan, Haryana, Punjab and Gujarat and coastal salinity caused by excessive withdrawal of ground water in Tamil Nadu, Saurashtra are other soil and water related constraints.

3.5 CLIMATE VARIABILITY

Climate change and climate variability are yet another emerging issue with serious implications on agriculture productivity. Global predictions suggest that the temperatures may rise by 0.6 to 2.5 by 2050 and 1.4 to 5.8 by 2100 (IPCC, 2007). In India, over the past 100 years, a temperature rise of 0.57° has been recorded (Rupakumar *et al.*, 1994). Though, so far no clear trend in the total rainfall has been noted, extreme events are on the rise. Various models predict that crop yields in India may decline in the range of 10 to 40% by the end of this century due to climate change, if no mitigation measures are taken.

More than total rainfall, the distribution is more important, particularly for rainfed crops. Long dry spells have significant negative impact on fodder and grain production indirectly affecting the livestock sector. Extreme weather events such as cold waves, heat waves, floods and high intensity single day rainfall events were on increasing trend from the last decade. For example, the 2002 drought across the country during kharif, the heat wave of May 2003 in AP, extreme cold winter in North during 2002-03, prolonged dry spell during July in 2004, abnormal temperatures during March, 2004 and January, 2005 in North, floods during 2005, cold wave during 2005-06, unusual floods in Rajasthan desert and drought in North-East in 2006 and abnormal temperatures during January-February, 2007 in North are some of the extreme weather events which had significantly impacted agriculture. Most models predict that these extreme events will further increase in future with significant impacts on livelihoods of the people.

3.5.1 Trends in Key Weather Parameters

Rainfall is the key variable influencing crop productivity in rainfed farming. Intermittent and prolonged droughts are a major cause of yield reduction in most crops. Long term data for India indicates that rainfed areas witness 3–4 drought years in every 10-year period. Of these, 2–3 are in moderate and one may be of severe intensity. However, so far no definite trend is seen on the frequency of droughts as a result of climate change. For any R&D and policy initiatives, it is important to know the spatial distribution of drought events in the country. A long term analysis of rainfall trends in India (1901 to 2004) using Mann Kendall test of significance by CRIDA indicate significant increase in rainfall trends in West Bengal, Central India, coastal regions, south western Andhra Pradesh and central Tamil Nadu. Significant decreasing trend was observed in central part of Jammu Kashmir, northern MP, Central and western part of UP, northern and central part of Chattisgarh (Fig. 3.6). Analysis of number of rainy days based on the IMD grid data from 1957 to 2007 showed declining trends in Chattisgarh, Madhya Pradesh, and Jammu Kashmir. In Chattisgarh and eastern Madhya

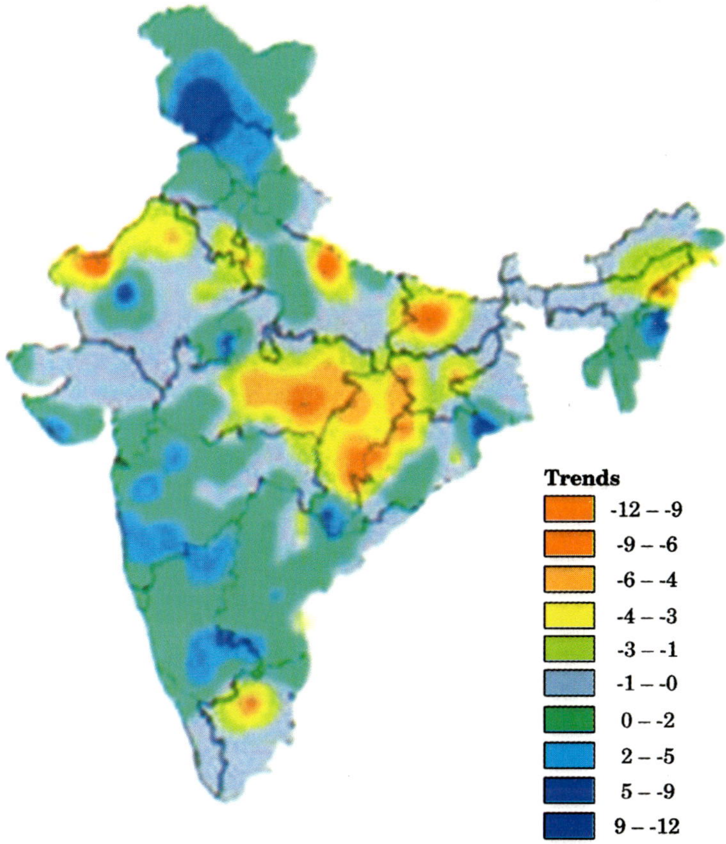

Fig. 3.6: Rainfall trends over India from 1901 to 2004. (*Source*: Data Bank, CRIDA, Hyderabad)

Pradesh, both rainfall and number of rainy days are declining which is a cause of concern as this is a rainfed rice production system supporting large tribal population who have poor coping capabilities.

Temperature is another important variable influencing crop production particularly during *rabi* season. A general warming trend has been predicted for India but knowing temporal and spatial distribution of the trend is of equal importance. An analysis carried out by All India Coordinated Research Project on Agrometeorology (AICPRAM), CRIDA using maximum and minimum temperature data for 47 stations across India showed that 9 out of 12 locations in south zone showed an increasing trend for maximum temperature, where as the north, only 20% locations showed increasing trend (Fig. 3.7). For

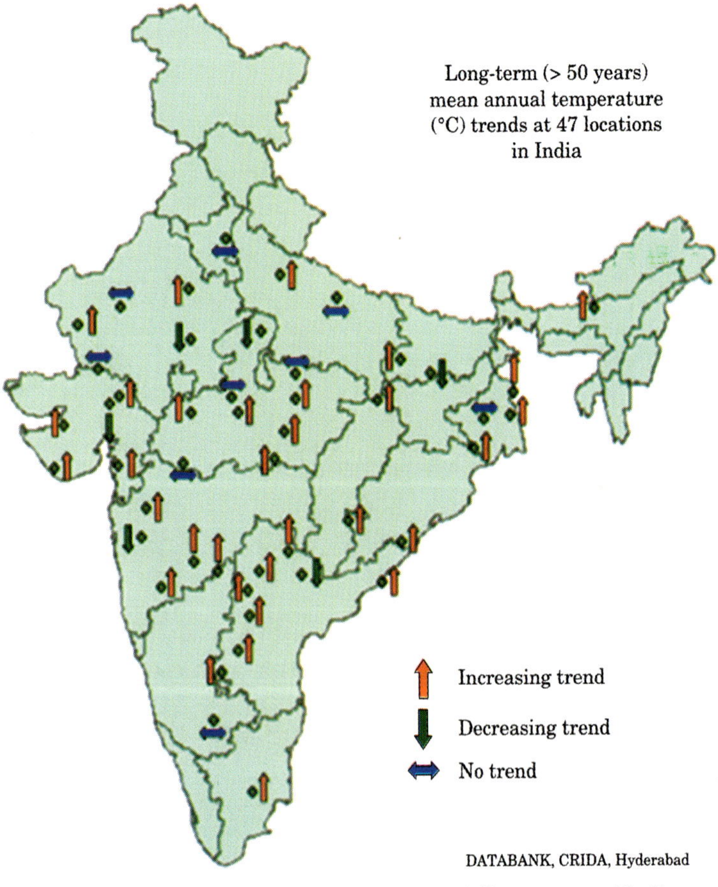

Fig. 3.7: Trends in mean temperature over different parts of India

minimum temperature most of the stations in India are showing increasing trend. This is a cause of concern for agriculture as increased night temperatures accelerate respiration, hasten crop maturity and reduce yields. The increasing trend is more evident in central and eastern zones where rainfall is also showing a declining trend which is an area of concern and requires high attention for adaptation research.

3.5.2 Impact on Water Requirements

Besides hastening crop maturity and reducing crop yields, increased temperatures will also increase the crop water requirement. A study carried out by CRIDA (unpublished) on the major crop growing districts in the country for four crops, *viz.*, groundnut, mustard, wheat and maize indicated a 3% increase in crop water requirement by 2020 and 7% by 2050 across all the crops/locations. The climate scenarios for 2020 and 2050 were obtained from HadCM3 model outputs using 1960–1990 as base line weather data (Table 3.5).

Table 3.5: Estimated crop water requirement (mm) of four crops in major growing districts of the country under climate change scenario

District (State)	*1990*	*2020*	*2050*	*% change over 1990 in*	
				2020	*2050*
Groundnut					
Tiruvanamalai (TN)	506.0	515.2	544.0	1.8	7.5
Rajkot (Gujarat)	559.3	562.3	582.2	0.5	4.1
Junagadh (Gujarat)	522.6	528.5	550.2	1.1	5.3
Belgaum (Karnataka)	354.9	366.3	386.0	3.2	8.7
Anantapur (AP)	517.6	567.1	650.3	9.6	25.6
Bangalore (Karnataka)	490.9	510.9	559.3	4.1	13.9
Mustard					
Agra (UP	276.4	284.0	295.5	2.7	6.9
Bharatpur (Raj)	276.2	283.8	295.1	2.8	6.8
Hisar (Haryana)	357.0	369.6	380.8	3.5	6.7
Nadia (WB)	483.2	491.5	508.8	1.7	5.3
Morena (MP	263.4	269.0	282.5	2.1	7.2
Wheat					
Sirsa (Haryana)	281.8	293.1	301.4	4.0	7.0
Ahmedabad (Gujarat)	523.0	536.8	551.0	2.6	5.4
Ahemedanagar (Mah)	485.8	496.1	509.5	2.1	4.9
Ganganagar (Raj)	278.9	290.3	298.2	4.1	6.9

Table 3.5: *(Contd...)*

Table 3.5: *(Contd...)*

District (State)	*1990*	*2020*	*2050*	*% change over 1990 in*	
				2020	*2050*
Hardoi (UP)	475.0	488.2	502.2	2.8	5.7
Kangra (HP)	367.7	380.7	391.2	3.5	6.4
Vidisha (MP)	437.1	446.9	460.4	2.3	5.3
Sangrur (Punjab)	391.1	405.4	416.3	3.7	6.4
Maize					
Udaipur (Raj)	388.8	392.4	400.9	0.9	3.1
Karimnagar (AP)	424.7	433.4	440.0	2.0	3.6
Jhabua (MP)	424.5	430.6	441.9	1.4	4.1
Begusarai (Bihar)	370.0	374.7	388.9	1.3	5.1
Bahraich (UP)	407.4	412.1	426.5	1.1	4.7
Godhra (Gujarat)	426.3	432.3	444.0	1.4	4.2
Khargaon (MP)	354.3	365.0	381.0	3.0	7.6
Aurangabad (Mah)	413.4	423.1	435.7	2.3	5.4

Source: Venkateswarlu (2010)

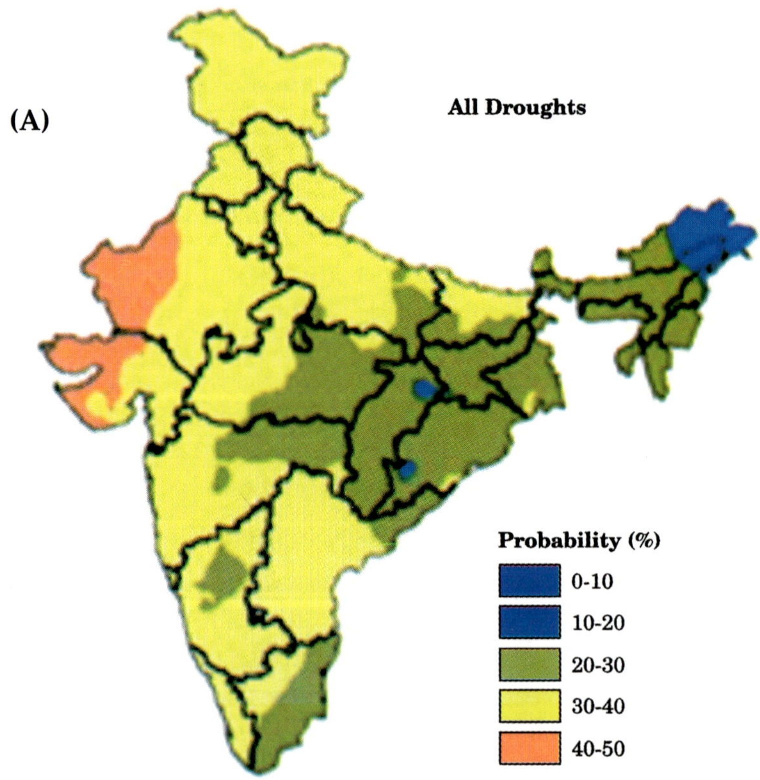

Source: Agromed databank, CRIDA, Hyderabad

Fig. 3.8a: *(Contd...)*

3.5.3. Greater Vulnerability of Rainfed Agriculture

While climate change impacts agriculture sector in general, rainfed agriculture is likely to be more vulnerable in view of its high dependency on monsoon, the likelihood of increased extreme weather events due to aberrant behavior of south west monsoon. Nearly 80 m ha of India's 141 m ha net sown area is rainfed. Rainfed farming area falls mainly in arid, semi-arid and dry sub-humid zones. About 74% of annual rainfall occurs during southwest monsoon (June to September). This rainfall exhibits high coefficient of variation particularly in arid and dry semi-arid regions. Skewed distribution has now become more common with reduction on number of rainy days. Aberrations in South-West monsoon which include delay in onset, long dry spells and early withdrawal, all of which affect the crops, strongly influence the productivity levels (Lal, 2001). These aberrations are likely to further increase in future. The risk of crop failure and poor yields always

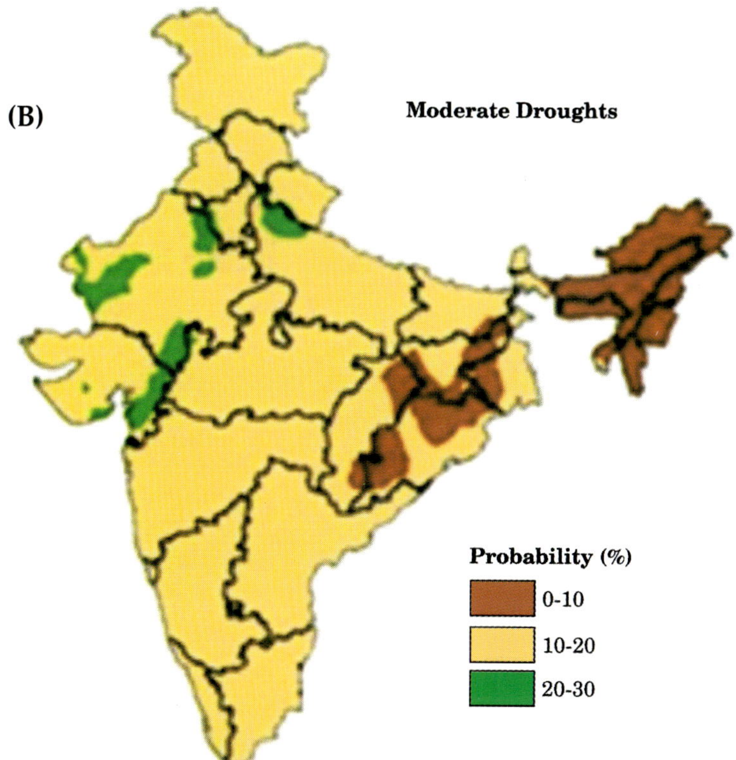

Fig. 3.8b: *(Contd...)*

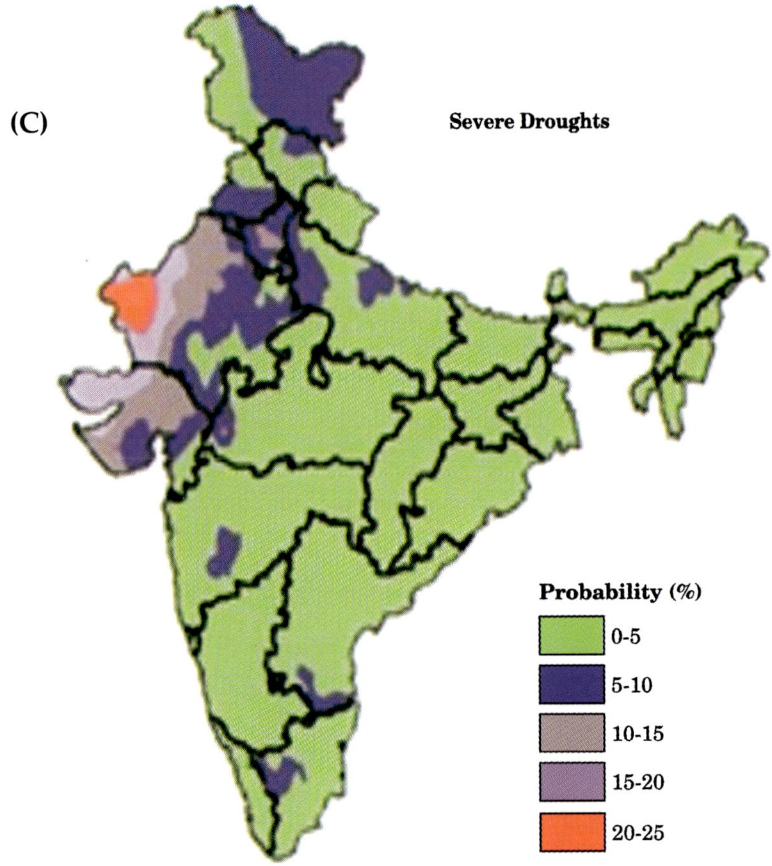

Fig. 3.8: Occurrence of different kinds of droughts in India (*Source*: Rao *et al.*, 2009) (A,B,C)

influence farmers' decision on investing on new technologies and level of input use (Pandey *et al.*, 2000). Numerous technological (*e.g.* cropping patterns, crop diversification, and shifts to drought-/salt-resistant varieties) and socio-economic (*e.g.* ownership of assets, access to services, and infrastructural support) factors will come into play in enhancing or constraining the current capacity of rainfed farmers to cope with climate change.

4

RESOURCE MANAGEMENT TECHNOLOGIES FOR SUSTAINING PRODUCTIVITY

Considering the importance of natural resources in sustaining agricultural productivity, extensive research efforts have gone into this area which resulted in development of efficient resource management technologies for different agro ecological regions. These include, better land use planning, soil fertility management and reclamation of problem soils, soil conservation, water harvesting, integrated watershed management, efficient use of irrigation water, etc. More recently, research attention was focused on conservation agriculture (CA) and some promising results are available from the rice-wheat cropping system in the Indo-Gangetic plains.

4.1 LAND USE PLANNING (LUP)

Land use planning (LUP) is the systematic assessment of physical, social and economic factors in such a way as to encourage and assist land users in selecting best land use options in an interactive and continuous process (FAO, 1993). Scientific land use planning is a pre-requisite for conservation of natural resources and sustaining agricultural productivity. Land use not supported by capability of land leads to soil degradation and sub-optimum productivity. In view of the fragmented land holdings and crop and livestock production being practiced mainly as a source of livelihood, adoption of scientific land use has not been possible in countries like India. However, its advantages for ensuring long term sustainability and productivity are now being realized by planners, scientists and farmers. Land use planning also helps Government to allocate land to not only to different sectors of the economy but various production systems within agriculture in order to meet the overall economic goals of the country.

NBSSLUP has delineated the country into 20 agro ecological regions (AER's) and 60 sub-regions (AESRs) for agricultural planning. Spatial variability in factors like bioclimate, physiography, soils and length of growing period were used in delineating the regions and sub regions with details on agro ecological settings and the land use potentials and constraints of each sub region along with information on major bench-mark soil series occurring in the sub region. AESRs are homogeneous land units to implement a wide range of land resources applications. The Agroecological Zoning of land resources appraisal linked to Geographic Information System (GIS) greatly enhanced the capability to develop alternate land use scenarios, management and decision support systems and improved interfaces to promote use of such systems by scientists, development administrators and land users (Velayutham *et al.*, 1999). Detailed soil site suitability criteria for different annual and perennial crops have been developed which helps in development of regional crop planning and operational plans for rehabilitation of degraded lands.

4.1.1 Scope of LUP

The advantages of scientific land use planning were evaluated at several locations on watershed basis under National Agricultural Technology Project by CRIDA and NBSSLUP. The results clearly brought out that land use based on the biophysical evaluation (soil resource inventory) and socio-economic assessment (PRA) could enhance 20–50% higher productivity of crops and livestock as compared to traditional land use followed by the farmers. The advantages were realized across the soil types like entisols, inceptisols, vertisols, alfisols and aridisols in arid, semiarid and sub-humid climatic zones. In most cases, the improved land use involved change of a crop or cropping pattern and minor change in the management practices addressing the specific constraints of the parcel of the land. These studies however brought out the fact that land use planning need to be done in a participatory manner considering both bio-phyiscal and socio-economic criteria. While the productivity and profitability are of immediate concern, the long term goal remains conservation of natural resources. By introducing the concept of land management units (LMU), it is possible to improve the productive capacity of soils by appropriate reclamation measures under National Rural Employment Guarantee Scheme (Ravindra Chary, 2005).

4.1.2 Land Capability and LUP

It is important to integrate the land capability class and rainfall for suggesting appropriate land use in a given region. Studies carried out under All India Coordinated Research Projects over 30 years have resulted in identification of optimum land use patterns for a variety of land capability classes under varying rainfall situations (Fig. 4.1). As the rainfall increases, the land use changes from pastures to crops, horticulture, paddy and fisheries in good quality lands. While, in poor quality lands, it moves from silvipasture to hortipasture and diary (Vittal *et al.*, 2006). Similar models are available for irrigated production systems.

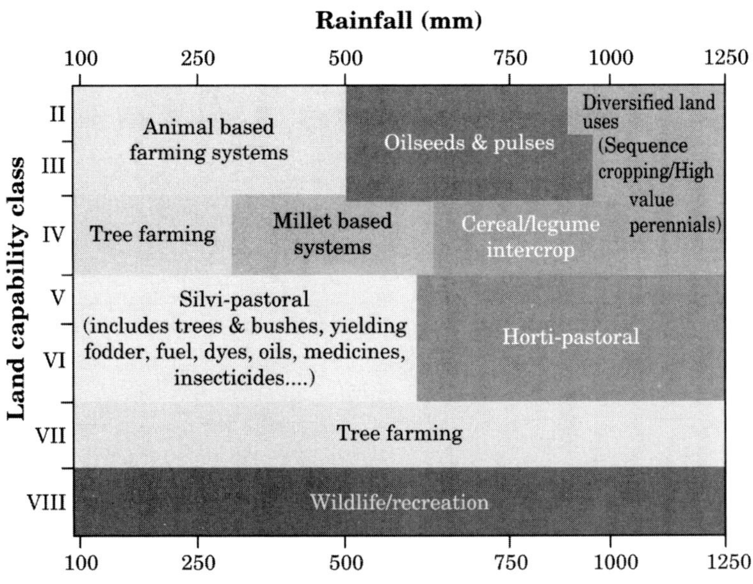

Fig. 4.1: Land use planning based on rainfall and land capability

With availability of satellite technology and geo-spatial mapping, it is possible to operationalise scientific land use planning to ensure sustainability of agriculture and livestock productivity. These technologies are to be used more widely at the state level planning. Land use planning also helps in drawing of plans for land reclamation, development of water resources and delineation of specific commodity wise production zones which will help the local Government to develop the required processing and marketing infrastructure.

4.2 SOIL FERTILITY MANAGEMENT

The single most important factor for increasing agricultural production in India in the next 25 years will be greater and more efficient use of nutrients, water and energy. To achieve targeted food production through intensification of agriculture, soil fertility and environmental safety are two factors, which need to be addressed. Higher yields per unit area and time means more removal of nutrients from soil and severe strain on the soil system. Some crops may remove particular nutrients more than the others and if fertilizer applications are not proportional to that, there will be net deficits and further imbalances. The net result would be a decline in the yield, in spite of heavy inputs.

4.2.1 Role of Fertilizers

Soil organic matter, being the storehouse of many plant nutrients has a significant effect on productivity. However, due to various constraints

Fig. 4.2: Cereal-legume rotation: A key principle in long term sustenance of soil fertility

discussed in earlier sections, application of organic manure and crop residues has declined in the country over time. Considering the growing nutrient imbalance, even if we use all organic manures available in the country, we will still have large deficit. Use of chemical fertilizers therefore is inevitable to sustain agricultural productivity. However, huge gap exists between consumption and production of N and P fertilizers. Potassium fertilizers are completely imported in India (Fig. 4.3).

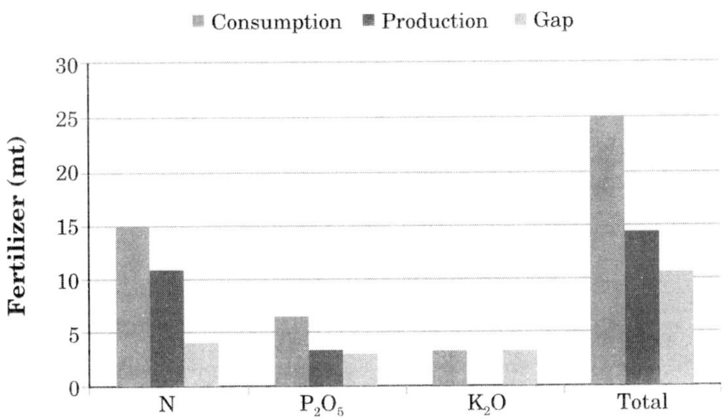

Fig. 4.3: Production and consumption of fertilizer nutrients in India (2008-09)

Fertilizers contribute about 50% of the increased yields as a component of improved technology. The dramatic increases in the yields of crops like wheat and rice have occurred because of high yielding varieties and higher fertilizer use. The yield potential of many dry land crops has not been tapped much despite introduction of good cultivars because of low fertilizer use. About 80% of the fertilizer is consumed in irrigated areas while only 20% is used in rainfed areas that constitute 60% of the cropped area (Venkateswarlu, 1986). Production and consumption of fertilizer nutrients in the year 2008-09 has been shown in Figure 4.3. The average fertilizer use in the country as a whole varied between 90-128.6 kg ha^{-1} in recent years, which is very low when compared with neighboring countries, like China (277.7 kg ha^{-1}), Japan (290.6 kg ha^{-1}) and Korea Republic (409.7 kg ha^{-1}). At the same time, the yield levels of crops like paddy, wheat maize, etc., in India are significantly lower compared to these countries. Hence, apart from many other reasons, low fertilizer use in India is definitely one of the important causes of low yields Srinivasarao *et al.* (2006, 2008, 2009a,c). Efforts need to be made to redefine the fertilizer doses by synchronizing with the anticipated water availability (Srinivasarao *et al.*, 2003). Out of total nutrient consumption, food grains share was 72 per cent followed by oil seed crops (9%) (Fig. 4.4) (FAI, 2009). Among food grains, paddy and wheat are major consumers of fertilizer nutrients (Fig. 4.5). Other crops like potato and coconut also have higher nutrient consumptions (Fig. 4.6).

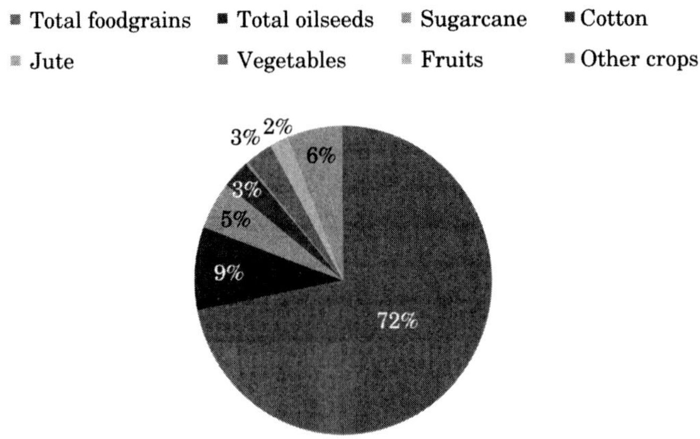

Fig. 4.4: Share of usage of fertilizer nutrients by various crops (2009)

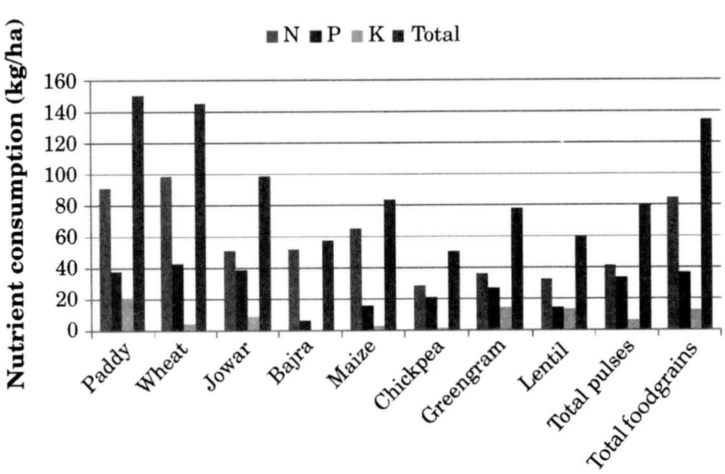

Fig. 4.5: Consumption of fertilizer nutrients by foodgrain crops

4.2.2 Crop Response to Fertilizers

The response of crops to fertilizer application varies depending on the production system. Long-term fertilizer experiments in alkali soils of the Indo-Gangetic plains showed that N and P were indispensable for sustaining the rice-wheat system, whereas pearl millet in the rotation did not respond to P application even when Olsen's P fell below critical level. In acid soils in the absence of liming, crop yields were negligible even at recommended levels of NPK due to severe acidification and

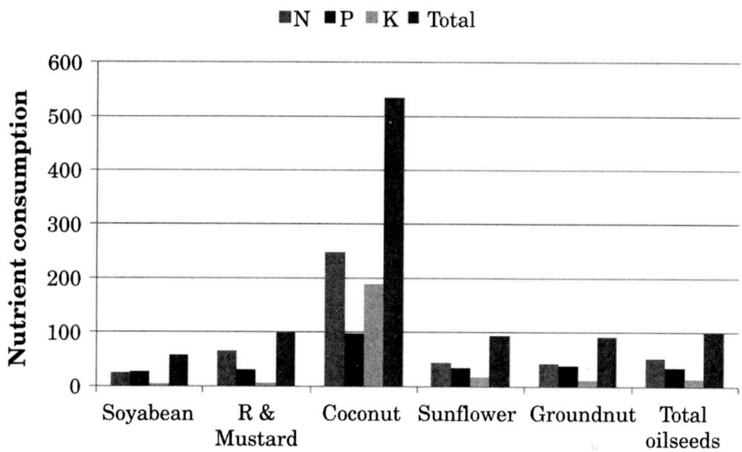

Fig. 4.6: Consumption of fertilizer nutrients by oilseed crops

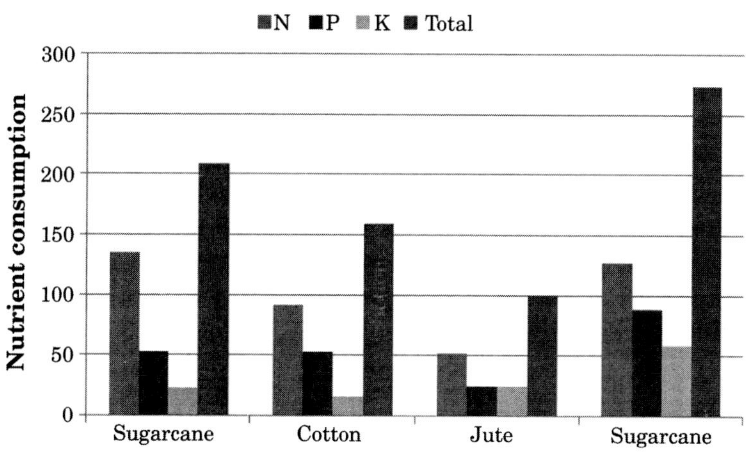

Fig. 4.7: Consumption of fertilizer nutrients by some high value crops

aluminium toxicity. The yields could only be sustained with application of lime or farmyard manure. On alkali soils, *Sesbania aculeata* green manuring or 15 t/ha of FYM besides 12.5 t/ha of gypsum (50% GR) was extremely important in maximising and sustaining rice and wheat yields and improving soil fertility (Swarup, 1998). In alluvial soils of Indo Gagetic plains, the response to N and P is declining due to the emergence of micronutrient deficiency and fall in organic matter levels.

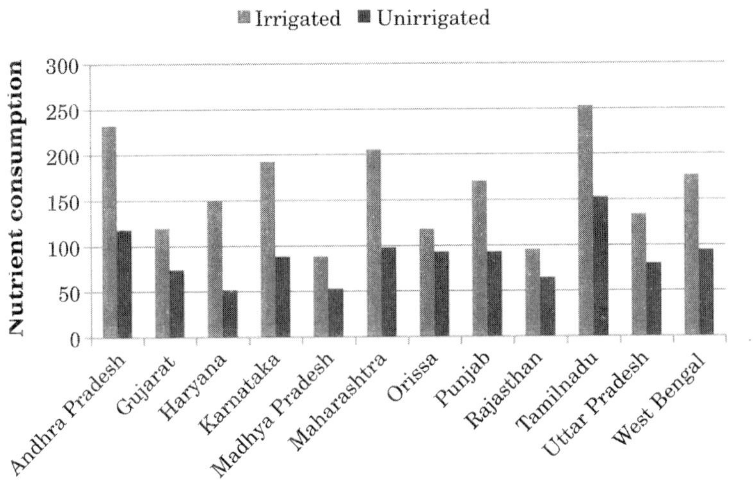

Fig. 4.8: State wise consumption of fertilizers under irrigated and unirrigated conditions

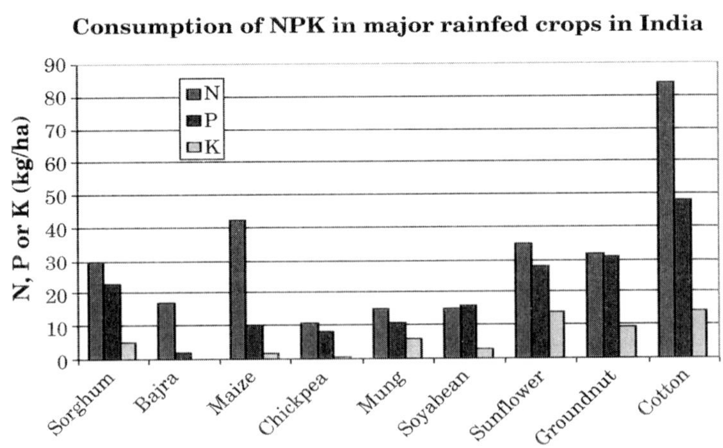

Fig. 4.9: Consumption of NPK in major rainfed crops in India

Extensive data were generated under the All India Coordinated Research Project for Dryland Agriculture (AICRPDA) on response of rainy and post-rainy crops to N and P application (Katyal *et al.*, 1997: Singh *et al.*, 1998: Katyal *et al.*, 1999a, Katyal *et al.*, 1999 b). For both the nutrients, the response varied from as low of 3 kg grain/kg nutrient to as high as 50 kg grain/kg nutrient. Unlike irrigated farming, the high variability in response to N and P, reflect the wide variations in crops and growing conditions between seasons and locations. Variation of 25–30% in response of the same crop between locations was observed (Srinivasarao *et al.*, 2011a). The relatively better recoveries of fertilizer N by crops (approximately 55%) support high returns from fertilizer N use with rainfed crops. Rao and Das (1982) reported that the response of rainy season sorghum to applied N varied from 3.4–43.4 kg grain kg^{-1} N but it was higher with post-rainy season crop at Bellary (6.5 kg grain kg^{-1} N) with 500 mm rainfall; Bijapur (9.7 kg) with 680 mm rainfall at Solapur with 722 mm rainfall and Kovilpatti (27.7 kg) with 700 mm rainfall. This response pattern also established a relationship with rainfall. Despite the proven benefits of fertilizer application, a number of constraints limit their widespread use in rainfed agriculture. The uncertainty of the onset and progress of the monsoon and socio-economic factors like infrastructure, credit and market access are the key constraints.

4.2.3 Enhancing Use Efficiency

Soil depth, moderates the influence of rainfall on fertilizer N use and losses. When water storage ability of the soil is less, the response is also less. Various factors influence fertilizer use efficiency in field conditions. Nitrogen use efficiency is influenced by the fertilizer, rate and time of application, and management factors such as tillage/crop variety, timely sowing, plant population, weed management and water management. To get the best out of N fertilizers, balanced nutrition approach is a must. Results from upland crops showed no significant difference between ammonium and nitrate containing N fertilizers in terms of their efficiencies. However, for rice, ammonical fertilizers were superior to nitrate-containing sources. The response to applied nitrogen by sesame was more with urea as compared to calcium ammonium nitrate and ammonium sulphate. In drylands, the best time and method of application is more important than the type of fertilizer (Srinivasarao *et al.*, 2009d). Tandon (1993), Srinivasarao *et al.* (2011a,b) comprehensively reviewed the strategies for efficient fertilizer management under dryland conditions.

Fig. 4.10: Crop response to site specific nutrient management in cotton (Khammam) and urdbean (Adilabad) in Andhra Pradesh

Method of application of fertilizers is also important. Placement of nitrogen (5–8 cm below surface) is more effective for rice crop. For dry land crops, separate application of seeds and fertilizers helps in avoiding direct contact between them which otherwise in adverse weather condition result in seedling mortality. Deep placement (10–15 cm below seed) is important for post-rainy season rainfed crops Srinivasarao, 2011). Yield and nitrogen recovery by safflower increased considerably by deep placement of nitrogen. Under rainfed conditions, the standard recommendation is to drill or place the basal dose few centimetres deep in the root zone and not to broadcast. In *kharif*, a part of the nitrogen plus all other nutrients are applied basally. In *rabi*, where little or no rainfall is expected, almost full dose of nutrients for the season is recommended basally. Efficient use of fertilizers have been reported by applying the fertilizers by band placement method (Venkateswarlu, 1987). The yield gains vary from 340 to 1500 kg grain/ ha. Foliar application of N is useful for late application in dryland crops, on sodic soils, and in submerged rice fields where soil application is not possible.

Apart from the fertilizer source and the method of application, the time of application of fertilizers is very important. Split (2 to 3) application of N matching the N requirement of the crop was found useful for most of the crops grown under irrigated conditions and under high rainfall situations. Under dryland conditions, since the amount and distribution of rainfall is erratic, fertilizer application has to match with the rainfall events. Two to three splits are recommended depending on the crop duration and the rainfall pattern. In a year of normal or relatively higher rainfall, little higher doses can be applied, and in a drought year, the dose can be reduced (Katyal *et al.*, 1991) saving resources to farmers. Efficiency of applied nutrients can be substantially improved by integrated nutrient management (INM). Based on 32 years long term manurial trail at Bangalore under rainfed condition, it was concluded that partial factor productivity can be sustained only with (INM) and organic manures while applying purely chemical fertilizers alone reduced PFP gradually in finger millet as depicted in Fig. 4.11 (Srinivasarao *et al.*, 2011b). Therefore, (INM) is essential for sustainability of agricultural production in India due to degraded soils with low organic matter and multi-nutrient deficiencies Srinivasarao *et al.* (2009c, 2011c, 2012a,b).

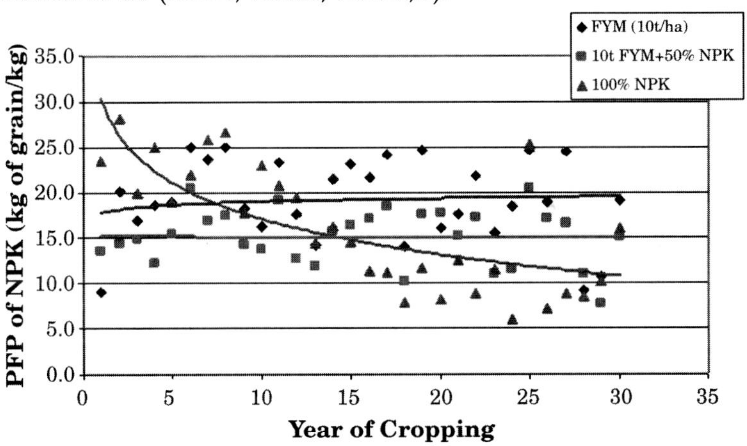

Fig. 4.11: Partial factor productivity of NPK in finger millet with various nutrient management options under rainfed conditions during 1978–2007 (Srinivasarao *et al.*, 2011b).

4.2.4 Water and Nutrient Interactions

The importance of fertilizer use in enhancing water use efficiency (WUE) has been well established in rainfed crops (Hegde *et al.*,1978). Application of fertilizers enables deeper root penetration and increased water extraction from soil. Application of P increased WUE by 15–20% in dryland wheat, 22–55% in finger millet, 41–99% in chickpea,

17% in linseed and up to 19% in mixed wheat-chickpea crop (Tandon, 1987). The increase in water use efficiency is more in coarse textured soils than in heavy soils. As the moisture becomes limiting (Srinivasarao *et al.*, 2009b,a), fertilizer application is to be optimised based on availability of water. Soil moisture and fertilizer interaction are quite significant in dryland crops (Singh and Das, 1995). Higher yield of post monsoon sorghum was obtained under deep soils as compared to shallow soils due to higher moisture storage capacity in the former (110 mm as compared to 50 mm in shallow soils). For instance, sorghum responded up to 25 kg N ha^{-1} in shallow soil and up to 50 kg ha^{-1} in deep soil. Similarly, higher responses to applied nitrogen were obtained in sandy loam soils as compared to loamy sand soils in wheat and maize as test crops. Moisture conservation practices and supplemental irrigation also enhance FUE.

Table 4.1: Response of rabi sorghum to nitrogen at different soil moisture availability on a vertic inceptisol

Soil depth (m)	*Available moisture storage (mm)*	*Grain yield (t/ha)N (kg/ha)*				*SE*	*CD*
		0	*25*	*50*	*75*		
0–30	50	0.66	1.23	1.30	1.30	0.111	0.327
0–60	110	1.66	1.78	2.13	2.33	0.120	0.333

Source: Randhawa and Singh (1983)

Table 4.2: Variations of crop response to applied fertilizer as influenced by rainfall distribution

Soil type	*Crop*	*Rainfall pattern*	*Yield without fertilizer (t/ha)*		*Kg grain/kg nutrient*	
			No of trials	*Yield (t/ha)*	*At 20–40 (kg/ha)*	*At 60–80 (kg/ha)*
Nitrogen						
Aridisols	Pearl-millet	Above normal	(3)	1.09	20.8	–
		Normal	(3)	0.75	13.7	–
		Below normal	(1)	0.25	6.3	–
Alfisols	Rice	Above normal	(2)	1.35	18.1	16.0
		Below normal	(3)	1.57	22.1	18.3
	Barley	Above normal	(1)	0.62	6.3	5.3
		Below normal	(2)	0.44	9.2	6.9
Inceptisols	Pearl-millet	Above normal	(2)	2.31	7.3	–
		Below normal	(2)	0.77	6.4	4.9

Table 4.2: (*Contd...*)

Table 4.2: (*Contd...*)

Soil type	*Crop*	*Rainfall pattern*	*Yield without fertilizer (t/ha)*		*Kg grain/kg nutrient*	
			No of trials	*Yield (t/ha)*	*At 20–40 (kg/ha)*	*At 60–80 (kg/ha)*
Entisols	Barley	Normal	(1)	2.50	10.0	11.8
		Below normal	(2)	1.29	14.8	13.7
	Wheat	Above normal	(1)	0.87	15.3	11.9
		Normal	(1)	0.72	3.2	0.5
Vertic	Sor-	Above normal	(1)	1.84	36.1	29.6
Inceptisols	ghum	Below normal	(1)	0.28	13.5	07.4
Phosphorus						
Aridisols	Green-	Above normal	(1)	1.32	2.7	1.8
	gram	Normal	(1)	0.30	1.0	1.3
		Below normal	(1)	0.12	2.7	1.8
Vertic	Sor-	Above normal	(1)	2.87	13.0	–
Inceptisols	ghum	Below normal	(2)	1.33	4.5	–
Inceptisols	Maize	Above normal	(1)	2.90	40.5	11.3
		Normal	(1)	1.93	7.2	–0.1

Source: AICRPDA Annual Reports

4.2.5 Integrated Plant Nutrient Supply (IPNS) and Management

Despite the proven benefits of chemical fertilizers, the falling response curves across the country due to emergence of micro and secondary nutrient deficiencies is a cause of concern. To address this, IPNS strategy has been suggested under different production systems. Data from long-term fertilizer experiments (LTFE) in India revealed that additional yields in different crops could be realized if only 10-15 tonnes of FYM was annually supplemented along with NPK (Table 4.3). Since increased agricultural production has to come from the existing lands only, soil fertility has to be sustained through continuous application of fertilizers and organics. Data from LTFE clearly indicate that with combined application of NPK and FYM, the latter contributes 20% to the total production (Swarup, 1998). A long-term experiment since 1972 in rice-wheat-jute system in West Bengal indicated that the balanced fertilization along with organic manures improved soil aggregation, biological activity and sustained productivity (Chaudhury *et al.*, 2005). Based on participatory soil fertility management trials in 90 villages in 8 districts of Andhra Pradesh, it was demonstrated that combined use of chemical fertilizers and locally available organic

Table 4.3: Average grain yields of crops (t ha^{-1}) over the years in long-term experiments across the country

Centre	*Crops*	*Control*	*N*	*NP*	*NPK 100%*	*NPK 150%*	*NPK + FYM 100%*	*NPK-S/ + S/lime*	*NPK + Zn*
Inceptisols									
Barrackpore	Rice	1.6	3.5	3.9	4.0	4.4	4.2	3.1	3.5
(1972–97)	Wheat	0.8	2.1	2.3	2.4	3.0	2.5	2.4	2.4
Bhubaneswar	Rice	1.6	2.1	2.2	2.8	3.0	3.5	2.4	2.8
(1973–94)	Rice	1.4	2.1	2.8	3.0	3.3	3.7	3.0	3.1
Coimbatore	Finger millet	0.7	1.0	2.8	2.8	3.0	3.3	2.8	2.8
(1972–95)	Maize	0.6	0.8	2.7	2.9	3.1	3.4	2.9	3.0
Delhi	Maize	1.2	1.6	1.8	2.1	2.4	2.3	2.1	2.2
(1994–96)	Wheat	2.2	3.5	3.9	4.3	4.9	4.6	4.8	4.8
Hyderabad	Rice	1.1	2.6	3.4	3.6	4.3	4.7	2.8	3.0
(1972–95)	Rice	1.0	1.8	2.2	2.6	3.3	3.4	2.3	2.5
Ludhiana	Maize	0.4	1.4	1.8	2.3	2.4	3.2	2.4	2.7
(1972–95)	Wheat	1.0	2.7	4.1	4.8	4.9	4.9	4.9	4.9
Vertisols									
Jabalpur	Soybean	1.0	1.2	2.0	2.2	2.2	2.3	2.0	2.0
(1972–96)	Wheat	1.1	1.4	4.0	4.3	4.5	4.7	4.3	4.0
Mollisols									
Pantnagar	Rice	3.4	5.0	5.0	5.4	5.4	6.2	5.7	5.2
(1972–96)	Wheat	1.6	3.8	3.8	3.9	4.2	4.6	4.2	3.8
Alfisols									
Palampur	Rice	0.3	0.6	2.0	3.2	4.0	4.6	4.0	3.6
(1973–97)	Wheat	0.3	0.4	1.8	2.5	3.0	3.3	3.1	2.5
Ranchi	Soybean	0.8	0.4	1.0	1.6	1.5	1.9	1.9	–
(1973–97)	Wheat	0.9	0.5	2.4	2.8	2.9	3.1	3.0	–
Bangalore	Finger millet	0.6	0.9	1.4	4.0	4.8	4.6	3.9	4.0
(1986–95)	Maize	0.4	0.7	1.0	2.3	2.5	2.7	2.3	2.2

Source: Swarup (1998)

resources like FYM, crop residue, glyricidia leaf manuring, Diancha green manuring, vermicompost, poultry manure, sheep penning, groundnut shells etc improved soil health, crop yields, yield stability, livelihoods of tribal population of Andhra Pradesh. Combined use of organic manures and chemical fertilizers enhanced the tolerance to water stress conditions during intermitted droughts in cotton (Srinivasarao *et al.*, 2011a) (Figs. 4.12, 4.13).

As stated earlier, a decline in productivity of rice-wheat rotation is observed in north and north-western Indo-Gangetic plains despite the use of optimum fertilizer inputs under assured irrigation. This is attributed to declining soil organic matter. Therefore, supplementary use of organics along with chemical fertilizers is inevitable (Swarup, 1998). Biofertilizers are another source. Benefits due to bio-fertilizers to irrigated and rainfed crops are well documented (Subba Rao, 2007;

Fig. 4.12: Glyricidia – Greenleaf manuring in Castor

Fig. 4.13: Horsegram as a cover crop to improve soil health and water retention in different agro-climatic regions of India (Arid Alfisol, left; Semi-arid Alfisols, right)

Venkateswarlu and Wani, 1999). Inoculation with *Azotobacter* and *Azospirillum* in rice and wheat gave 8–10% yield increase, and higher increases of 10–20% in millets, sunflower, etc. Use of phosphate solubilizers, Vescicular Arbuscular Micorrhiza (VAM), blue green algae (BGA) and other N fixers have also shown promising benefits (Rao and Patra, 2009; Venkateswarlu, 2007). Green manuring with *Sesbania rostrata, Sesbania aculeata* and green gram in rice-wheat system not only improved soil organic matter but improved the physical properties. Mandal *et al.* (2003) recorded a reduction in bulk density due to green manuring over fallow in 0–15 cm and 15–30 cm soil layers during the growth of rice and wheat. Green manuring also improved soil aggregation as evident from higher values of mean weight diameter and saturated hydraulic conductivity than fallow.

IPNS is also relevant for rainfed agriculture (Katyal *et al.*, 1999a; Sharma *et al.*, 1999, 2008). The practice of effective use of inorganic and organic sources of nutrients in a proper proportion not only reduces the requirement of inorganic fertilizers but also improves physical conditions of soil and enhances soil water retention Srinivasarao *et al.* (2011, 2012a,b). Grain yields of finger millet with optimum NPK application were similar to those obtained from 50% NPK and 10 t/ha FYM in an Alfisol at Bangalore (Hegde *et al.*, 1988). A three years study on an Alfisol at Hyderabad revealed that conjunctive use of organic sources of N such as loppings + twigs of N fixing trees like *Gliricidia maculata* or *Leucaena leucocephala* + urea in 1:1 ratio (equivalent to 40 and 80 kg N/ha) gave 17.2 and 16.9 q/ha yield of grain sorghum, respectively which were equal to that obtained with 100% N applied through urea alone. The apparent recovery of N applied was more at 41.5% with conjunctive application (Sharma and Srinivas, 1997). There are several other reports of better performance of fertilizers when used in combination with organic sources. The predominant INM recommendations for rainfed crops are listed in Table 4.4.

The advantages of IPNS approach in arid region is also well document (Rao and Singh, 1993). Aggarwal and Praveen Kumar (1993), on the basis of six year long study reported the direct beneficial effect of FYM as well as its synergistic effect on crop yield when applied in combination with inorganic fertilizers. This practice also helped in improving N use efficiency of urea and fertility status of soil. Other studies on nutrient management in arid regions (rainfed) clearly indicate a positive effect of organic manures, legume based crop rotation and crop residue incorporation in maintaining soil fertility for a sustainable crop production.

Table 4.4: Effective Integrated Nutrient Management (INM) practices for rainfed crops across the country

Location		*Fertilizer (kg ha^{-1})*			
		N	*P_2O_5*	*K_2O*	
Jhansi	Cluster bean	15	60	0	Inoculation with
	Sorghum+Dolichos	60	20	0	*Rhizobium*
Rajkot	Sorghum	90	30	0	FYM@ 6 t ha^{-1}
	Pearlmillet	80	40	0	FYM@ 6 t ha^{-1}
	Groundnut	12	25	0	FYM@ 6 t ha^{-1}
	Cotton	40	0	0	FYM@ 6 t ha^{-1}
Solapur	Sorghum	50	0	0	9–10t/ha subabul loppings can substitute 25 kg N/ha^{-1}
Indore	General	(N plus P)			4–6 t ha^{-1} FYM
	Soybean	20	13	0	in alternate years FYM@ 6 t ha^{-1}
Bijapur		(NP or NPK)			Mulching with tree lopping @ 5 t ha^{-1}
Arjia	Corn–pigeonpea	50	30	0	50% N through organics.
	Safflower and rapeseed-mustard	30	15	0	Reduction in N by half if these crops follow legumes such as greengram/ chickpea
Agra	Barley	60	30	0	Use of FYM plus *Azotobacter*
Ranchi	Soybean	20	80	40	Inoculation with *Rhizobium*
	Groundnut	25	50	20	Inoculation with *Rhizobium*
	Pulses	20	40	0	Inoculation with *Rhizobium*
Dantiwada	Greengram	0	20	0	Inoculation with R*hizobium*
Jodhpur	Pearlmillet	10	0	0	Addition of 10 t/ha FYM
Hoshiarpur	Corn	80	40	20	Addition of FYM
	Wheat	80	40	0	
	Chickpea	15	40	0	
Akola	Cotton+greengram	25	25	0	Along with FYM to meet 25 kg N ha^{-1}

Source: AICRPDA Annual Reports

IPNS approach not only helps in improving the crop yield by providing the primary nutrients but also supplies micro and secondary nutrients. LTFE results indicate that at recommended levels of NPK without Zn, the productivity of cropping systems remained high for about 10–12 years but declined subsequently due to increasing Zn deficiency. In the LTFE initiated at Pantnagar in 1971, Ram (1998) observed that over a 25-year period, DTPA–extractable Zn in soil showed a declining trend till 1993, but it improved when it was applied to soil or partially restored with INM (Table 4.5). Deficiency of Fe can also be corrected through organic matter addition as evident from the significant increase in available Fe status of soil with the regular incorporation of Sesbania green manure before transplanting rice in the coarse textured soils (Singh *et al.*, 2005). Green manuring of rice also increased the DTPA extractable Mn due to the conversion of crystalline oxide bound Mn into easily reducible amorphous forms in the reduced conditions produced by the incorporation of green manure.

Table 4.5: Status of available micronutrients (mg/kg) after 25 years of continuous rice-cowpea fodder cropping at Pantnagar

Treatment	*Zn*	*Fe*	*Mn*	*Cu*
100% NPK	0.77	38.0	24.7	4.2
100% NPK + Zn	2.29	35.0	23.7	4.3
100% NPK + FYM	1.12	44.5	30.0	5.7
150% NPK	0.60	41.3	29.0	4.5
Initial value	2.70	29.5	26.8	2.9

Source: Ram (1998)

4.2.6 Site-specific Nutrient Management (SSNM) and Precision Farming

SSNM is an approach of feeding the crops with external nutrients, based on the initial soil fertility status, type of crop and variety, desired yield and nutrient uptake pattern. SSNM has been evaluated in rice throughout China, India and South East Asia (Dobermann *et al.*, 2004). Compared with traditional local management, SSNM demonstrated 7, 30 and 12% increase in yield, recovery efficiency of N (REN), and net returns, respectively. More systematic research is needed to capitalize on the SSNM approach in India. In Karnataka, nutrient application on the basis of SSNM principles resulted in significantly higher grain yields over FP (farmers' practice) and RDF (recommended fertilizer dose) in case of rice, wheat and chickpea (Table 4.6) as reported by Biradar *et al.* (2006).

Precision farming can also help in improving the use efficiency of inputs and enhancing yields. It requires knowledge of what inputs are required, where and when. Advantage of precision agriculture is derived when the resulting information is turned into management decisions. Fertilizer use based on yield potential associated with specific soil factors such as water holding capacity are other approaches that would optimize yield.

Table 4.6: Yield of wheat, rice, and chickpea (t ha^{-1}) as influenced by site specific nutrient management (SSNM) in Karnataka

Site	Wheat			Rice			Chickpea		
	SSNM	RDF	FP	SSNM	RDF	FP	SSNM	RDF	FP
1	3.70	3.20 (16)[1]	2.70 (37)[1]	5.70	4.20 (36)	3.70 (54)	1.38	1.14 (21)	1.13 (22)
2	3.80	2.84 (34)	2.60 (46)	5.32	4.00 (33)	3.56 (49)	1.18	1.03 (15)	1.01 (17)
3	3.50	2.96 (18)	2.70 (30)	5.50	4.06 (36)	3.91 (41)	1.22	1.08 (13)	1.08 (13)
4	3.60	3.00 (20)	2.64 (36)	5.00	3.71 (35)	3.36 (49)	1.25	1.06 (18)	1.05 (19)
5	3.72	2.90 (28)	2.56 (45)	6.08	4.50 (35)	3.90 (56)	1.26	1.07 (18)	1.06 (19)
Mean	3.66	2.98 (23)	2.64 (35)	5.52	4.09 (35)	3.69 (50)	1.26	1.08 (17)	1.06 (18)

Numbers in parenthesis indicate % increase in yield under SSNM over RDF or FP.
Source: Biradar *et al.*, 2006

The future nutrient management strategies have to be more crop and soil specific with aim to achieve highest use efficiency. Preventing environmental losses will become a high priority. SSNM, precision farming, use of customized fertilizers, tailoring nutrient management with conservation agriculture and minimizing green house gas emissions through better fertilizer application are some of the priorities which need attention (Srinivasarao *et al.*, 2011d).

Nutrient deficiencies scenario at different research stations in the IGP under intensive rice-wheat cropping system has been depicted in Table 4.7. In the soil test based approach fertilizer rates are established based on the concept of crop removal, with an adjustment for soil residual nutrients. While this approach actually fits most production systems in India quite well, given that most of the crop biomass is removed from harvested fields, the role that residual soil nutrients play in meeting crop nutrient requirements becomes a challenge. For example, if a soil tests medium or low in most plant nutrients, then application of these nutrients based on crop removal from a target yield is going to address these nutrient demands. However, on soils where the soil nutrient analysis indicates a high level of nutrient supply, the issue of whether to apply the nutrient at removal rates becomes a challenge to the researcher. The best option, therefore, is to apply all macro and secondary nutrients required to meet crop yield removal and those micronutrients that soil testing show to be deficient. This then provides the environment for full yield expression in absence of any nutrient deficiency. And once this yield potential of a site has been determined, the next step is to refine nutrient application rates with further yield

Table 4.7: Nutrient deficiencies observed at different research stations in the IGP under rice-wheat cropping system

Centre	*Nutrient deficient*							
	P	*K*	*S*	*Zn*	*Fe*	*Mn*	*Cu*	*B*
PDCSR, Modipuram	–	√	√	√	–	√	√	√
GBPUA & T, Pantnagar	√	√	–	√	–	√	–	√
CSAUA & T, Kanpur	√	√	√	√	–	–	–	–
NDUA & T, Faizabad	√	√	√	√	–	√	–	√
BHU, Varanasi	√	√	√	√	–	√	√	√
RAU, Sabour	√	√	√	–	–	–	–	–
BAU, Ranchi	√	√	√	√	–	–	–	√
HPKV, Palampur	√	√	√	√	–	–	–	√
PAU, Ludhiana	√	√	√	√	√	√	√	
RS Pura	√	√	√	√	–	√	√	–

Better Crops (IPNI), 2010.

trials. The positive impact of this approach to fertilization was clearly shown in a series of research experiments on soil-test based SSNM in rice-rice and rice-wheat cropping systems in seven different locations in the IGP. When the yield-limiting nutrients were identified and applied at each location as a SSNM treatment, it was able to generate large improvements in yield and profitability over farm practice across all sites. A smaller gap existed between SSNM and the State recommendation, although most sites still suggested an economic advantage for the SSNM approach (Table 4.8).

The SSNM approach was evaluated comprehensively for agronomic, economic and environmental performance in 56 farmer's fields with irrigated wheat and transplanted rice in Punjab. The results of the study clearly brought out the positive impact of SSNM on grain

Table 4.8: Effect of site-specific nutrient management (SSNM) on wheat productivity (t/ha) and economic return (INR/ha) in parentheses at seven locations in India

Site	*Farm practice*	*State recommen-dation*	*SSNM*	*Increase over SR, % (INR/ha)*	*Increase over FP, % (INR/ha)*
Ranchi	2.56 (1,575)	4.15 (25,276)	4.06 (26,854)	–2.2 (1,578)	58.5 (25,309)
Modipuram	4.77 (29,292)	4.90 (31,859)	6.43 (58,083)	31.0 (26,224)	46.5 (28,791)
Kanpur	4.72 (7,258)	5.45 (17,644)	6.00 (31,338)	10.1 (13,694)	27.1 (24,080)
Ludhiana	5.45 (27,772)	6.28 (39,105)	6.55 (46,219)	4.3 (7,114)	20.1 (18,447)
Sabour	3.92 (18,306)	4.97 (28,614)	5.82 (45,116)	17.1 (16,502)	48.7 (26,810)
Pantnagar	3.87 (7,828)	5.10 (14,276)	6.39 (19,426)	25.3 (5,150)	66.0 (11,598)
Palampur	2.64 (55,122)	3.76 (54,583)	3.87 (60,905)	3.0 (6,322)	46.5 (5,783)

Better Crops (IPNI), 2010.

yields, and agronomic, recovery and physiological efficiencies of N under rice-wheat cropping system in Punjab *vis-a-vis* farmer practice (Table 4.9).

Table 4.9: Grain yield of rice and wheat, agronomic (AEN), recovery (REN), and physiological (PEN) N efficiencies, total fertilizer cost (TFC), and gross returns above fertilizer cost (GRF) in 56 farmer fields under rice-wheat cropping system in Punjab

	Rice		*Wheat*	
	FFP	*SSNM*	*FFP*	*SSNM*
Grain yield, kg/ha	5.1	6	4.2	4.7
AEN, kg grain/kg N	8.8	16.1	8.3	13.6
REN, kg N/100 kg N	20	30	17	27
PEN, kg grain/kg N	34.7	44.2	29.4	37.1
TFC (INR/10 ha)	23,055	34,930	31,059	34,800
GRF (INR/ha)	24,578	28,014	22,316	25,274

4.3 MANAGEMENT OF PROBLEM SOILS

The major problem soils in India include those affected by salinity and alkalinity, soil acidity and water logging. Problem soils occur in different parts of the country, both due to natural and manmade factors. Substantial improvements in productivity have been demonstrated by management of such soils with appropriate and cost effective technologies developed over time.

4.3.1 Reclamation of Salt Affected Soils

Reclamation of saline and sodic soils was practiced in India much before the beginning of formal research. Any reclamation technique to be successful with saline and alkaline soils, three essential requirements have to be met, *viz*. (i) salt or alkali must be completely removed from the root zone, (ii) the land must be prevented from reverting to the original condition and iii) the damage already done to the soil should be corrected (Kelly, 1951). Velayuthum and Bhattacharya (2000) have comprehensively discussed three broad methods to reclaim saline – alkali soils which could be grouped into (i) physical and hydro–technical amelioration; (ii) biological amelioration; and (iii) chemical amelioration. Under physical amelioration, the predominant treatments include: deep ploughing, sub-soiling, sand filling and profile inversion. The first three techniques improve water infiltration. Opening of land and breaking up the hard surface is possible through mechanical means. Sub-soiling consists of breaking the hard pan through the use of chisels drawn by powerful tractors, provided the pan lies within 40 cm to 100 cm of the surface. Sand filling or 'sanding' is done to improve water and air permeability. After suitable mechanical treatment of the soil, hydro-technical procedure is employed to leach the salts from root zone by irrigation or rainfall.

Under biological method of amelioration, addition of organic matter benefits saline and alkali soils principally in two ways; (a) improvement in water infiltration, and (b) release of carbon dioxide during decomposition. Even growing a crop has biological ameliorative action, since during respiration CO_2 is produced and plants exert shading effect reducing evaporation and slowing down upward rise of salts from ground waters. In calcareous soils, the CO_2 brings in soluble calcium for exchange reaction to replace sodium in the exchange complex. Biological activity can be stimulated by allowing a grass cover to develop; or through afforestation, or green manuring with suitable crops; or incorporation of crop residues, application of farmyard manure, pressmud and other organic materials. Since, partially composted organic materials release much more CO_2 than the thoroughly composted (digested) organic wastes; the former extends more calcium solubilizing action than the latter.

Feed back from farmers indicate sodic and saline-sodic soils amelioration is preferred through 'Vegetative bioremediation' (Kumar and Abrol, 1984; Robbins, 1986a; Quadir *et al.*, 2002; Mishra *et al.*, 2002). Vegetative bioremediation of sodic and saline sodic soils is achieved by the ability of plant roots to increase the dissolution rate of calcite, thereby resulting in enhanced levels of Ca^{+2} in soil solution. The vegetative bioremediation increases the partial pressure of CO_2 within the root zone, enhance the proton release in the root zone in case of certain N fixing crops, improves soil aggregation as well as hydraulic properties, and some plant roots remove sodium from the soil. The collective effect of these factors ultimately leads to soil amelioration provided leaching and drainage are adequate. Inorganic chemicals are used to reclaim alkali soils, which can supply either directly or indirectly, calcium that can replace exchangeable sodium. These agents are: (a) calcium chloride or gypsum; (b) acidifying materials like sulphur, iron sulphate, iron pyrites, aluminium sulphate, etc. Of these, gypsum is by far, the most widely used amendment. Other waste materials like phosphogypsum, fly ash, magnesite etc. are finding use to certain extent, in reclamation of salt-affected soils but their systematic evaluation is necessary before large-scale application is recommended. In actual practice, the reclamation can be done more effectively by combining various ameliorative methods, since interaction between them brings in more spectacular results. The synergic effect of the combination of methods is well documented (Gupta and Abrol, 1990).

The evaluation of vegetative and chemical approaches in different countries reveals comparable performance of both in terms of sodic soil amelioration. Results of a field experiment (Kumar and Abrol, 1984) conducted on a barren calcareous alkaline soil (pH = 10.2, EC = 2.7 dSm^{-1}, ESP = 94) indicated that the amelioration efficiency of two grasses, para grass (*Brachairia mutica* (Forssk.) Stapf) and Karnal grass was comparable with soil application of gypsum at 12.5 Mg ha^{-1}. Ahmad *et al.*, (1990) compared sesbania, kallar grass and sordan as vegetative bioremediation treatments against gypsum application (13 Mg ha^{-1}) on a calcareous, sandy clay loam, saline-sodic field. The treatments effectiveness was in order: gypsum = sesbania > kallar grass > millet rice > finger millet. Forage yields of the plant species were found directly proportional to their amelioration efficiency.

4.3.2 Management of Acid Soils

Acidity of soil creates unfavorable medium for soil micro flora in decomposition of organic matter. It also makes primary, secondary and micronutrients to remain fixed or insoluble form. "Phosphate fixation" in highly acid soils is an acute problem. The phosphate gets fixed with soluble iron present in acid soils. High rainfall with high temperature and heavy leaching are the main factor for the formation of acid soils. Liming is recommended when the soil is highly acidic and where multicropping involving acid sensitive crops is adopted. Liming improves base saturation and availability of Ca and Mg. Fixation of P and Mo is reduced by inactivating the reactive constituents. Toxicity arising from excess soluble Al, Fe and Mn is corrected and thereby root growth is promoted and uptake of nutrients is improved. Liming also stimulates microbial activity and encourages N_2 fixation and nitrogen mineralization, and hence, legumes are highly benefited from liming. The liming material may be ground lime stone ($CaCO_3$), burnt lime (CaO) or hydrated lime Ca $(OH)_{2-}$ When lime is added, the soil solution becomes charged with calcium ions which replace hydrogen ions in the exchange complex. When burnt (CaO) or slaked or hydrated lime $Ca(OH)_2$ is added to acidic soil, it changes into calcium bicarbonate, which in solution reacts with soil colloids. The highly acid soils exist in the states of Kerala, Chattisgadh and West Bengal and North East states (Fig. 4.14).

In liming programme, the first and foremost thing is to assess the lime requirement of the soil for optimum yield of the crop/crops. The desirable dose of lime for laterite soils was however found to be half of the lime requirement (LR) dose determined by Woodruffs buffer (pH 7.0) for maize and cotton. Economic response was recorded to lime sludge applied at 0.5 LR dose in a variety of crops in red loam soils with pH 5.0. When applied in high doses, much of the lime is lost by leaching from the topsoil of light textured soils because of their low exchange capacity. Split application is recommended to minimize the leaching loss. Lime should preferably be applied in smaller doses more frequently (every alternate year) rather than in heavy dose once in three to four years. Application of full LR dose suppresses P availability and cause B deficiency. Lime should be applied before ploughing or on ploughed land well before cropping and should be thoroughly mixed with the soil by discing.

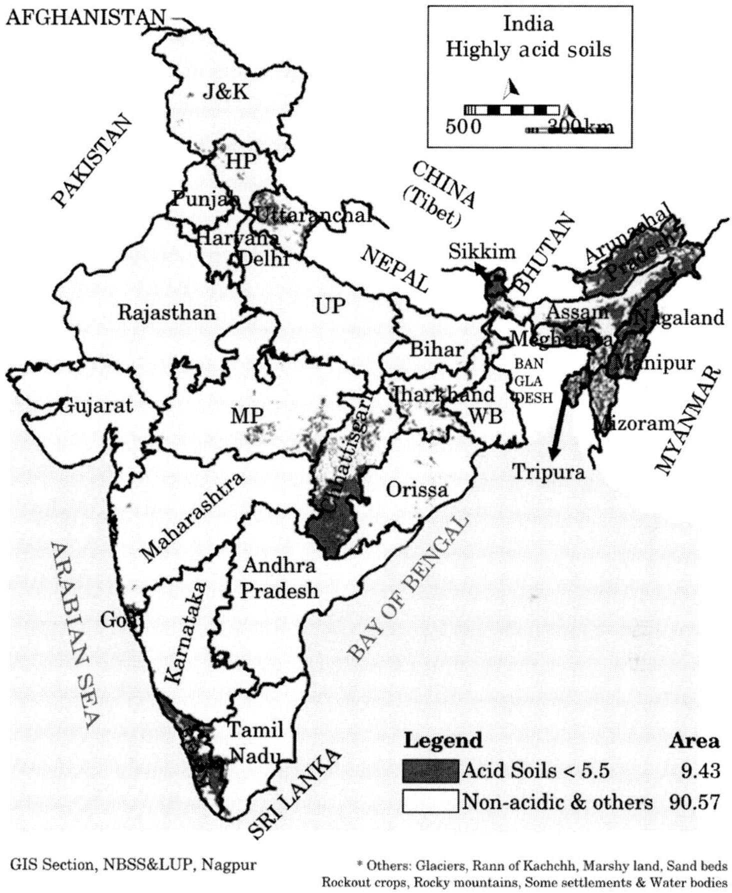

Fig. 4.14: Acid soils of India (*Source*: Maji *et al.*, 2008. *Annual Report of NBSS&LUP*)

The results from several long term fertilizer experiments conducted in different agro ecological regions involving diversified cropping systems and soil types have shown that imbalanced fertilizer use particularly N alone had a deleterious effect on soil productivity and health and the damaging effects in the absence of P and K fertilizers varied in the order Alfisols > Vertisols > Inceptisols > Mollisols. In a period of 24 years (1973–97) average productivity of soybean and wheat in N alone plots was lesser than control in Alfisols (acid soils) of Ranchi (Swarup, 2010) (Fig. 4.15).

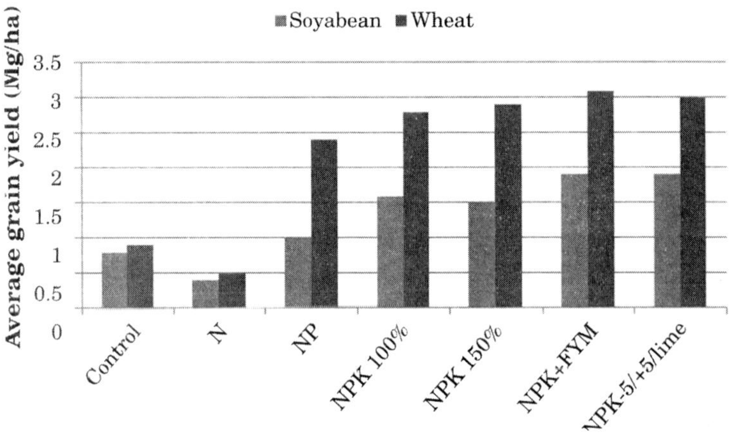

Fig. 4.15: Average grain yields of soybean and wheat with different treatments in acid soils of Ranchi

4.3.3 Management of Water Logged Soils

Waterlogging results in restriction of the normal circulation of air. The depth of water-table which is considered harmful depends upon the crop, type of soil and water quality. This depth may vary over a wide range from surface for rice to about 1.5 m for other crops. Of the total waterlogged area of 6 million ha, 3.4 million ha are subject to surface flooding in different states. Remedial measures used are: (a) Surface or sub-surface drainage to remove surplus water, (b) lining of canals to prevent seepage and rise of water-table, and (c) sinking tube well and utilizing water for irrigation thereby lowering water-table. Waterlogged lands contiguous to the sea-coast may be ameliorated for crop cultivation or used for fish culture (Anonymous, 1976). Prevention of waterlogging and secondary salinisation in the command areas of irrigation projects is as important as the need for erosion-control measures in the catchment areas. Thus, owing to injudicious soils and water management many irrigated areas are producing hardly one-half to one-third of what is otherwise possible (Yadav, 1984). This can be augmented by proper water management (Velayuthum and Bhattacharya, 2000).

4.3.3.1 *Managing sea water intrusion in coastal areas:* **Doruvu/Kottai** *technology*

Deep (upto 20 ft) open well is used compared to shallow ponds in traditional system. Horizontal flow of under ground water enabled in to the well through pipes which facilitates more water storage in lesser area. In this system more water can be pumped and more area can be covered under irrigation.

Traditional system

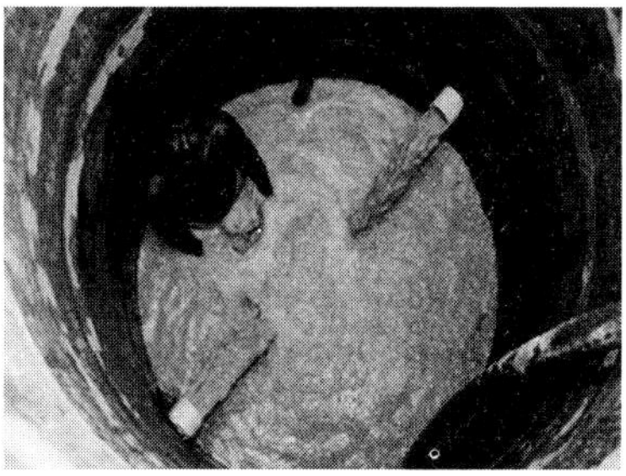

Improved system

4.4 SOIL CONSERVATION, WATER HARVESTING AND WATERSHED MANAGEMENT

Although India receives relatively high rainfall, there are large temporal and spatial variations. Unlike temperate countries, rainfall occurs in high intensity storms carrying away top soil through runoff. Because of the skewed distribution of the rainfall, crops invariably experience water stress. Therefore, soil and water conservation and harvesting of surplus runoff are of paramount importance. The total water received through annual precipitation in India is estimated at 400 m ha m. About 300 m ha m is received during June to September, while another 100 m ha m during the rest of the year. About 20 m ha m of runoff is brought from outside India, thus making a total of 420 m ha m of water resource for use. The cardinal principles of soil and water conservation are: (i) conserve rainfall by reducing runoff losses and by storing as much as possible in the profile and (ii) adopt farm practices to make the most efficient use of the soil moisture by crops (Gupta *et al.*, 2000).

In the beginning, physical structures like bunding were emphasized, but more recently the focus shifted to field- and community-based *in situ* conservation practices. The main aim of these practices is to reduce or prevent either water erosion or wind erosion, while providing the desired moisture for crop production. Based on past experiences, several field based soil and water conservation measures have been found promising for various rainfall zones in India (Table 4.10).

Table 4.10: Recommended soil and water conservation measures for various rainfall zones in India

Seasonal rainfall (mm)			
<500	*500–700*	*750–1000*	*>1000*
Contour cultivation with conservation furrows	Contour cultivation with conservation furrows	BBF(Vertisols) conservation furrows	BBF (Vertisols) field bunds
Ridging	Ridging	Sowing across	Vegetative
Sowing across slopes	Sowing across slopes	Slopes	Bunds
Mulching	Scoops	Tillage	Graded bunds
Scoops	Tide ridges	Lock and spill	Chos
Tied ridges	Mulching	Drains	Level terraces
Off-season tillage	Zingg terrace	Small basins	
Inter row water	Off-season tillage	Field bunds	
Harvesting system	BBF	Vegetative bunds	
Small basins	Inter row water	Graded bunds	
Contour bunds	Harvesting system	Nadi	
Field bunds	Small basins	Zingg terrace	
Khadin	Modified contour bunds		
	Field bunds		
	Khadin		

Source: Vittal *et al.* (2003)

4.4.1 Runoff Control

Runoff is inevitable when rainfall intensity exceeds the intake rate of the soil. A study by Mishra *et al.* (1999) revealed that in the red soils of Hyderabad, about 60 percent of the annual rainfall causes runoff. On an average, 12 major rain storms occur over a period of 5 to 7 months in a year. Conservation measures are effective in reducing the runoff to about 2 percent of the total rainfall. Various methods are available to trap the rainfall within cultivable area and to increase deep percolation. Some of the proven land treatments and their impacts on crop yields are presented in Table 4.11.

Table 4.11: Effect of land treatment on crop yield at different rainfed locations in India

Location	*Climate and soil*	*Annual average rainfall (mm)*	*Test crop*	*Land treatment*	*Increase in yield over control*
Hoshiarpur (Ballowal Saunkhri)	Sub-humid Inceptisol	900	Maize	Sowing across the slope	24
Anantapur	Arid Alfisols	560	Groundnut	Dead furrows at 3.6-m intervals	10
Rewa	Sub-humid Vertisols	1080	Chickpea	Channel after 5 rows	17
Hyderabad	Semi-arid Alfisols	760	Sorghum	Sowing across the slope and ridging later	19
Bhubaneswar	Sub-humid Oxisols	1500	Cowpea	Ridge-furrow cultivation	49
Solapur	Semi-arid Vertisols	722	*Rabi* sorghum	Compartmental bunding	25
Bellary	Semi-arid Vertisols	500	*Rabi* sorghum	Contour farming	35
Akola	Semi-arid Vertisols	830	Sorghum: Cotton	Broad-bed and furrows	26 37
Rajkot	Semi-arid Vertisols	625	Groundnut	Ridge and furrows	26
Bijapur	Semi-arid Vertisols	680	*Rabi* sorghum	Rubble check	47
Kovilpatti	Semi-arid Vertic Inceptisols	750	Pearlmillet	Compartmental bunding	23
Arjia	Semi-arid Vertisols	658	Maize	Ridge and furrow	35
Agra	Arid Inceptisols	710	Pearlmillet	Compartmental bunding	19

Source: Krishnappa *et al.*, (1999)

Even in areas where rainfall is ample, its uneven distribution affects crop yields due to excess water at one time and water stress at another. The effects of this skewed distribution have to be evened out. In some areas, it can be done by providing irrigation, in others, particularly sub-humid and humid regions, the surplus can be stored and utilized during deficit period.

4.4.2 Water Conservation

About 40 percent of annual precipitation goes as runoff and hence this precious resource is to be captured for crop production. Storage of water in the soil and in natural or man-made structures and efficient utilization of given quantity of water are important components of water harvesting (Gupta *et al.*, 2000; Singh, 2000). *In situ* water conservation is a more feasible and practical proposition under most situations but the constraints for *in situ* moisture conservation lie in soil management, which aims to maximize the use of rainfall by increasing infiltration and storage. Soil cover management (mulching/canopy), tillage and land configurations (ridge and furrow, BBF, etc.) are practices aimed at increasing infiltration and soil moisture storage.

4.4.2.1 *In situ water harvesting for increasing crop productivity*

Drylands often experience intermittent dry spells during cropping season causing moisture stress, particularly in low and medium rainfall areas. One way of overcoming this problem is to conserve as much rainfall *in situ*. Several technologies were evolved in the past to address this problem such as mulching, land configurations, application of tank silts, etc. However, the adoption of such practices is not satisfactory as farmers face constraints like higher cost of cultivation, weed problem, lack of suitable equipment, non availability of mulch material and most importantly, additional labour requirement.

One common *in situ* conservation practice recommended for Alfisols in Telengana region is conservation furrow. It is made after first weeding *i.e.* 30–45 DAS (days after sowing) so as to combine two field operations of weeding and making furrows. Though this conserves the moisture thereafter, the runoff causing rains received during first 30–45 DAS are not stored in profile and utilized for crop growth. This called for a technology which can merge well with existing farmers practice with out additional expenditure. This should also use entire seasonal rainfall for moisture conservation.

Thus a new technology of making of conservation furrow, seeding, fertilizer application all at planting time was tested by CRIDA, Hyderabad during *kharif* 2009. This involves paired row planting with ridge planter. It consists of a 25 cm deep of furrow with width of 60 cm at 120 cm interval and seeding with a seed drill on either side of furrows with required spacing. It was tested on pigeonpea, castor and cotton. With this practice, 250 m^3 harvesting capacity is created *in-situ* for conservation of rainfall thus alleviating the water need for measures to overcome a 15 days dryspell. A ridger was developed which can be attached to conventional seed cum fertilizer drill for making conservation furrows. This system also has the potential to overcome transient water logging in vertisols in high rainfall areas.

Sowing with ridge and furrow equipment in paired row with conservation furrows

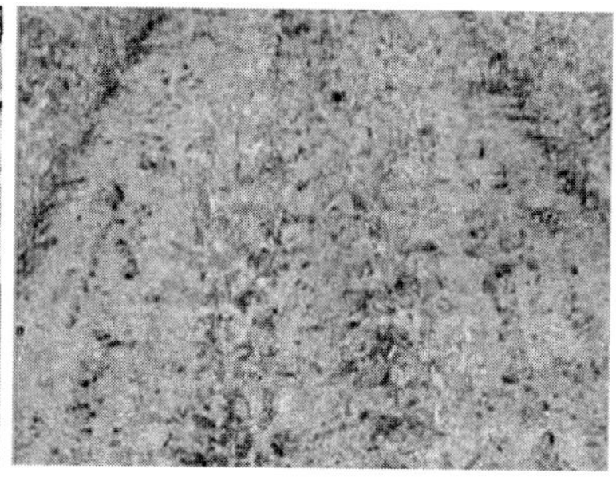

Pigeon pea - paired row planting

Building on *in situ* moisture reserves to tide over recurring drought spells, disallowing subsequent loss of profile moisture, permitting safe runoff disposal; its collection and recycling are three principal strategies. The individual strategy varies with the soil hydrologic characteristics and rainfall (AICRPDA 2003).

4.4.2.2 *Building in situ moisture reserves to tide over the recurring drought spells*

Off-season (summer tillage) coinciding with pre-monsoon showers helps in increasing rainwater infiltration, moisture conservation and efficient weed control. This results in increased crop yields in Alfisols with textural discontinuity. However, deep and tap rooted crops sown a few weeks after the onset of rainy season are not significantly benefited by this practice (Table 4.12). Land leveling in deep alluviums helped in increasing yields especially in sub-normal seasons (Table 4.13).

In chronic drought prone areas, deep tillage (20–30 cm) was found beneficial to soils having textural profiles or hard pans (Table 4.14). Under uni-modal (<500 mm) rainfall situation in semi arid regions with shallow Alfisols, sowing across the slope and ridging later was useful. It assisted in greater rainwater absorption and control of pernicious weeds.

Table 4.12: Effect of off-season tillage on the yield of sorghum (3 rainy seasons) and castorbean (2 rainy seasons) on Alfisols at Hyderabad and barley (3 post-rainy seasons) on Inceptisols at Varanasi

Practice	*Sorghum (t/ha)*	*Clusterbean (t/ha)*	*Barley (t/ha)*
Without off-season tillage (farmers practice)	1.87	0.32	1.37
With off-season tillage (improved system)	2.60	0.31	1.57

Table 4.13: Effect of land shaping on the yields of wheat and chickpea in sub-montane soils at Ballowal-Saunkhri and pearlmillet in alluvial soils of Agra

		Pearlmillet (t/ha)	
		Rainfall 320 mm	*Rainfall 660 mm*
Natural	1.05	1.34	2.22
With land leveling	1.65	1.63	2.31

Table 4.14: Crop response to primary deep tillage as drought proofing approach under three types of drought situations

	Grain (t/ha)		*% increase*
	Shallow tillage (10–15 cm)	*Deep tillage (20–30 cm)*	
Severe Drought			
Coarse cereals (5)	0.92	1.30	41
Legumes (6)	0.61	0.76	25
Moderate Drought			
Coarse cereals (16)	2.14	2.37	11
Legumes (2)	1.21	1.38	14
Less Prone to Drought			
Coarse cereals (4)	1.55	1.72	11
Legumes (1)	0.25	0.29	16

In black soils, deep tillage alone is of no avail. Tillage combined with compartmental bunding is an effective soil management practice for Vertic Inceptisols. With rainy season cropped Vertisols (unimodal rainfall regions), water surplussing is an integral part of *in situ* moisture conservation. In bi-modal medium (500–750 mm) rainfall regions representing semi-arid Alfisol, graded border strips are advantageous. Extensive tillage of sandy soils (common to desert areas) made them more vulnerable to wind erosion. Superficial scraping to eliminate weeds was found adequate.

4.4.2.3 *Preventing loss of stored soil moisture*

Tillage combined with seeding across the slope reduced runoff losses by 10 to 20%. Runoff thus averted could sustain crop growth for another 2 weeks following intermittent failure of rains. Effectiveness of cultivation across the slope significantly increased if post-planting intercultivation was directed to develop ridges and furrows between the crop rows. The yield increases attributable to these two practices varied between 30 and 40%. Ridges and furrows are required to be laid on a minor gradient of 0.3 to 0.4% to prevent temporary water logging. The ridge-and-furrow system is not practicable with closely spaced row-sown crops (row spacing <30 cm). In such instances, the alternative is creation of deep furrows (popular as 'dead furrows') after 8 to 10 crop rows (~3 m interval) for rainwater capture and disposal, if necessary.

In rainy season-cropped, unimodal low rainfall regions (mean annual rainfall 600–750 mm), ridges and furrows with water surplus were found useful. Sowing on flat without ridging later on as a part of cultural operations is a traditional practice. The serration of the soil surface is an efficient way of permitting more time for water entry into the soil profile. Ridges and furrows always increased the yield but its effectiveness is more so under moderate drought. In other situations like severe drought or in regions less prone to drought, the effectiveness is constrained by the lack of water or excess of water (Table 4.15).

In rainy season-cropped, unimodal medium to high rainfall regions (mean annual rainfall <800 mm), raised and sunken bed system

Table 4.15: Crop responses to land treatments under 3 types of droughts

	Grain (t/ha)		*% increase*
	Flat	*Ridges-Furrow system*	
Severe Drought			
Coarse cereals (19)	1.73	2.02	17
Legumes (24)	0.71	0.78	10
Moderate Drought			
Coarse cereals (12)	1.51	1.92	28
Legumes (1)	0.30	0.42	40
Less Prone to Drought			
Coarse cereals (1)	0.79	0.91	15

is most appropriate to provide drainage and storage of runoff. In post-rainy season-cropped bimodal low-rainfall regions (mean annual rainfall <750 mm), scooping, compartment bunding and tied ridging are effective for in *situ* water conservation. In Vertisols of medium depth, dividing the fields into sectors of 3 m × 3 m by compartment bunds led to yield increases up to 50%. In post-rainy season cropped heavy black soils (mean annual rainfall 550–600 mm), vertical mulch increased water intake and reduced runoff. Effect of vertical mulch in increasing yield was more visible during sub normal rainfall seasons (Table 4.16).

Table 4.16: Influence of vertical mulch on post-rainy season sorghum on deep vertisols

Practice (seasons)	*Grain (t/ha)*
Control	1.31
Vertical mulch at 5 m interval (4)	1.65
Vertical mulch at 10 m interval (4)	1.24

Mulching with *Leucaena* loppings and other green leaf manures helps in improving the soil properties. Use of organic mulches reduced evaporation losses of soil-stored water, increased stability and sustainability of crop yields particularly of post-rainy season crops (Table 4.17).

Table 4.17: Effect of mulching on amelioration of midseason drought

Crops (seasons)	*No mulch (t/ha)*	*Mulch (t/ha)*	*Response (%)*
Severe Drought			
Coarse cereals (5)	1.44	1.39	–4
Legumes (8)	0.55	0.70	27
Moderate Drought			
Coarse cereals (19)	1.40	1.81	29
Legumes (3)	0.29	0.33	13
Less Prone to Drought			
Coarse cereals (1)	1.04	1.44	38

4.4.2.4 *Permitting safe runoff disposal and its harvesting in dugouts*

Well established vegetative barriers, similar to mechanical bunds help in con-servation of soil and rain water. No distinct superiority of *Vetiveria zizoniodes* (Khus) a much talked about vegetative barrier could be established over *Cenchrus ciliaris, Cymbopogan slaxuosus, Pennisetum hohenackeri* and *Leucaena leucocephalan* both in red soils and black soils across the centers. In rainy season-cropped Vertisol medium rainfall regions, broad bed and furrow system, which combines benefits of good aeration and safe disposal of excess rainwater, is a useful technology. Water harvesting in small farm ponds (100–250 m^3) was developed as a viable technology. While the technology itself is viable and replicable high infiltration losses due to seepage and surface evaporation were the main constraints faced. Among the various sealants tested in Alfisol regions, brick and mortar lining was found to be most efficient and economical in the long run. Water resource thus developed, could be used for life saving irrigation, extension of cropping season, increasing cropping intensity and raising of high-value cash crops. Inter-plot water harvesting (1:1 cropped) is yet another simple technique which harvests rainwater *in situ*. It increased crop yields in Aridisols by 50% and Alfisols in dry sub humid regions by more than 17 per cent.

4.4.3 Recycling of Harvested Runoff

Trials on recycling of harvested water at different dryland centres proved the effectiveness of one supplemental irrigation of 5 cm in increasing the yield and water use efficiency (WUE) of several dryland crops (Table 4.18).

Table 4.18: Effect of one supplemental irrigation (5 cm) on yield of rainfed crops

		Yield (q/ha)		*% increase*
		Without irrigation	*With irrigation*	
Wheat	Agra	21.9	27.4	110
Barley	Varanasi	26.0	33.6	152
Sorghum	Bijapur	16.5	23.6	148
Upland rice	Rewa	16.2	27.8	232
Fingermillet	Bangalore	16.2	23.2	140
Safflower	Basad	15.6	20.2	92
Chickpea	Basad	10.8	16.3	110
Pearlmillet	Jodhpur	16.6	19.0	48

Source: Singh (1982)

4.4.4 Watershed Management

India has the world's largest area under irrigation (60 m ha). Still, 67% net sown area is rainfed. The world's largest quantity of groundwater

(200 cubic km) is pumped out annually in the country for agriculture. A huge investment of about 3 billion US dollars per year is made in watershed development alone for land and water management (www.mssrf.org). The watershed management programmes in the country have become pivotal for developing rainfed regions. The Planning Commission Working Group on Watershed Development, Rainfed Farming and NRM for the tenth Plan assessed that 88.5 million ha area of degraded wasteland (including rainfed areas) would need development. The Group report envisaged covering the entire 88.5 million ha area in four successive five year plans, commencing from the tenth plan up to the thirteenth plan at an estimated cost of Rs. 72,750 crore (on 1994 prices). Out of the total geographical area of the country, about 120 m ha is degraded and 85 m ha is rainfed cropped land. This includes degraded land not only under private ownership, but also with the departments of panchayat, revenue and forest. All these lands arc prioritizcd for development under various watershed development projects. The Government of India has established a National Rainfed Area Authority (NRAA) with an objective to coordinate and converge programmes in rainfed areas under different ministries.

Life saving irrigation of through harvested rainwater

4.4.5 Evolution of the Watershed Programme

Over a period of time, watershed program has evolved from purely technical orientation program to community oriented participatory strategy. In 1983, ICAR piloted a National Watershed Programme involving 47 model watersheds across the country through CRIDA, Hyderabad and CSWCR & TI, Dehradun. Launching of NWDPRA (National Watershed Development Programme for Rainfed Areas) has its roots in the successes achieved through these models. Central Ministries (MoA, MoRD, MoEF, etc.) and departments are implementing programmes. The watershed programme in eighties was mostly top driven with little or no stake holder's participation. During nineties, people's participation was emphasised but issues of usufruct rights, management of CPRs and conflicting norms of different ministries emerged. After 2000, many of these issues are resolved, and currently the programme is more oriented towards livelihood enhancement and convergence with other development programmes at the Panchayat level.

Fig. 4.16: Watershed activitites for water harvesting and efficient water utilization

4.4.6 Impact of Watershed Programmes

CRIDA carried out SWOT analysis of 37 watersheds under different agro-climatic regions and implementing agencies. Three types of survey was taken up in the study *viz.*, (i) rapid reconnaissance survey; (ii) rapid rural appraisal survey and (iii) detailed field survey. The selected physical, social and economic parameters were examined in detail. The highlights of the survey are:

1. The Watershed Development Programme (WDP) encouraged farmers to bring more area (9.0%) under plough.
2. The irrigated area increased from 38.2% in non-watershed areas to 52.4% in watershed area.
3. All the socio-economic indicators showed marked improvement in watershed over non-watershed areas. But, large and medium farmers (LMF) were more benefited from the programme (Rs. 4000/ha) because, the programme was land based.
4. LMF derived greater benefit even from livestock. Stall feeding was adopted more by LMF.

5. The Government funded programmes scored poorly in adopting participatory approaches in comparison to externally aided projects and the programmes of NGOs. People's participation (%) in planning, execution, monitoring and evaluation and maintenance of watersheds implemented by different agencies is given in the Table 4.19.

Table 4.19: Participation percentage of the primary stakeholders in planning, execution, monitoring and evaluation and maintenance components in watersheds (n = 37)

Agency/Indicators	*Participation(%) of the primary stakeholders*			
	Planning	*Execution*	*M&E*	*Maintenance*
NWDPRA	75	62	31	22
MoRD	82	75	40	41
ICAR	91	79	16	15
IA	92	90	20	37
NGO	90	86	36	36
Average	86	78	29	30

6. Training and capacity building require more attention in watersheds implemented by government agencies.
7. In respect of crop production, cereal production was more with small and marginal farmers. In other commodities, the difference between small/marginal farmers and large farmers was not significant.
8. The yield of selected crops between watershed and non-watershed areas is significant (Table 4.20). However, in the case of maize, green gram and groundnut, there were no differences.

Table 4.20: Yield of selected crops between watershed and non-watershed areas

	Average yield (q/ha)		*% increase*
	Watershed	*Non-watershed*	
Cereals			
Sorghum	11.2	9.4	19.1
Pearlmillet	12.2	9.4	29.8
Ragi	14.2	10.3	37.9
Maize	22.6	23.0	–1.7
Rice	22.0	20.3	15.0
Wheat	23.5	18.8	25.0
Oilseeds			
Groundnut	10.4	10.6	–1.9
Soyabean	14.2	6.2	129.0
Pulses			
Greengram	8.9	8.9	0.0
Blackgram	8.9	5.9	50.8

9. Milk production was generally more in the watershed areas.
10. Watershed areas practiced conservation better, evidently, due to the awareness created in the programme. They opted more for field leveling and field bunding than contour bunding. The over all improvements in physical parameters in watershed projects in various agro-climatic conditions are given in Table 4.21.

Table 4.21: Improvements in physical parameters in watershed projects in various agro-climatic conditions (n = 37)

Parameter/Agro-climate	*Arid*	*Semiarid*	*Humid*
Rise in water table (m)	1.05	1.57	1.38
Reduction in runoff (%)	35.0	33.2	30.5
Reduction in soil erosion (%)	15.0	28.8	25.6
Surface water resources developed (%)	9.0	18.0	20.5
Increase in afforestation (%)	10.0	11.3	21.7
Increase in cropping intensity (%)	6.0	16.0	18.3
Increase in employment (%)	12.5	25.0	20.8

11. The LMF practiced soil conservation better than the SMF both in and outside watershed areas.
12. *In situ* rain water harvesting was practiced more in watersheds. Of the various practices, summer ploughing and criss-cross ploughing were the most accepted.
13. Individual structures like farm ponds and wells and cascade of check dams for ground water recharge were the most preferred structures by farmers.
14. The rise in groundwater was more in watersheds and large & marginal farmers were more benefited from it.
15. Employment generation was more (7% more with male and 14% more with female) for agricultural workers. The women in the watershed as well as non-watersheds worked for 12–13 hours a day. However, there was a saving of 18% time in fetching water and fuel wood for household purposes in the watershed areas thus reducing drudgery.

A meta analysis of watershed programs by ICRISAT covering 311 watersheds pointed to better returns (IRR, BCR), equity and employment generation from watersheds in the rainfall region of 700–1100 mm (Joshi *et al.*, 2005). In general, the small and marginal farmers did not get the benefits to the desired extent. In terms of higher returns, livestock-pasture based land use was found ideal for regions receiving rainfall upto 700 mm, crop-horticulture for regions between 700–1100 mm and crop-fish based systems for regions receiving more than 1100 mm rainfall. However, location specific need based interventions, participation of the community are key to successful watershed programmes.

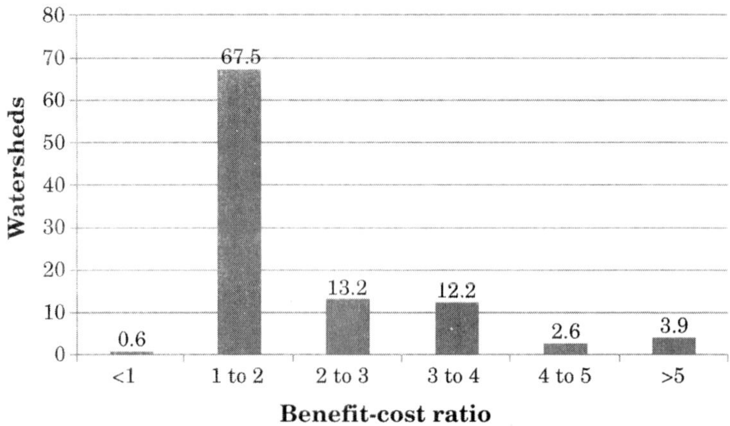

Fig. 4.17: Distribution (%) of watersheds according to BCR

4.5 WATER MANAGEMENT IN IRRIGATED AGRI-CULTURE

India receives average annual rainfall of 1194 mm with a variation of about 100 mm/year in Jaisalmair of Rajasthan to 11,690 mm/year in Mousinram, Cherapunji of north-eastern region of India. With 4000 billion cubic meter of annual rainfall, average runoff generated is only 1869 BCM. Due to various constraints about 1122 BCM of water can be put to beneficial use of which 690 BCM is through surface water and 432 BCM by groundwater. Out of 690 BCM of surface water, so far about 213 BCM of storage are built through major irrigation projects. Another 184 BCM of storage are under construction. Similarly out of 432 BCM of groundwater resource, about 360 BCM of groundwater is expected to be available for irrigation out of which present usage is about 135 BCM.

The spectacular growth in food production in India (from merely 50.8 million tonnes in 1950–51 to about 235.88 million tonnes in 2010–

11) is mainly attributable to three major factors; namely, development of new genotypes in cereal crops, large expansion in irrigated area and fertilizer use. Among these, irrigation may be considered as a backbone of modern intensive agriculture, as it not only ensures physiological water needs of crop plants for desired growth of photosynthetically efficient new genotypes but also supports efficient utilization of plant nutrients. But in spite of all these impressive achievements, net sown area in the country remain at about 142 Mha and net irrigated area is only 54.7 Mha. Out of net sown areas of 142 Mha, the ultimate irrigation potential is 139.9 Mha. But to meet the projected food grains demand of 494 million tonnes by 2050, it is required to bring 146 Mha areas under irrigation. This is quite high compared to the present gross irrigated area of around 99 Mha. Thus, it will be an enormous task to achieve the requirement of 146 Mha of irrigated area. On the other hand it is also essential to increase the crop productivity from present 2.3 tonnes/ha to 4 tonnes/ha under irrigated conditions and from 1 ton/ha to 1.5 tonnes/ha in rainfed area. Net area under irrigation is required to be increased to catch up with our increasing population for food security and to prevent the decline of share of agriculture to gross domestic products of the country.

Therefore, to produce more food with the same amount of water, productivity of water has to be improved dramatically and improved irrigation water management is an important step in this regard. Better timing of water supplies can reduce stress at critical stages of crop growth period and increase yields. With reliability in water supplies, the farmers can confidently invest on inputs for higher productivity. Supplementing rainfall to irrigation can increase productivity of water when limited supply is available to crops at critical periods. Precision irrigation using drip and sprinkler technology, proper conveyance system as well as better information and communication technology can reduce non beneficial evaporation, seepage and percolation losses, apply water uniformly to crop and reduce stress. Synergy among water management, high yielding variety, fertilizer and plant protection measures will result in quantum jump in productivity in agriculture.

With growing urbanization in the country, there will be more requirement of water from competing sectors like domestic and industrial needs. Per capita water resources of the country is also continuously declining from 5176 m^3 in 1951 to 4732 m^3 in 1955, 2209 m^3 in 1991, 1820 m^3 in 2001 and 1703.6 m^3 in 2005. The population of India is expected to stabilize around 1640 million by the year 2050. As a result the per capita water availability will further decline to 1140 m^3/year (Table 4.22). For adequate living standards as in western and industrialized countries, a renewable water supply of at least 2000 m^3 per capita per year is necessary (Postel, 1992). Hence, with per capita water availability of 1140 m^3/year, India will be water stressed.

Table 4.22: Per capita water availibility in India (past, present and future)

Year	*Population (in million)*	*Per capita water availibility (in cubic metre)*
1951	361	5177
1955	395	4732
1991	846	2209
2001	1027	1820
2025	1394 (projected)	1341
2050	1640 (projected)	1140

Source: Hand Book of Agroculture, 2009, ICAR, New Delhi

Country's water demand is also dominated by irrigation needs. Different demand scenarios upto 2050 AD are given in Table 4.23. This demand of a region depends upon the areas irrigated with surface water and ground water, different crop water requirement and the irrigation application efficiency.

Table 4.23: Future food grains and water demands for irrigation on under different scenarios

Year	*Low food grains demand (mt)*	*Demand water requirement (BCM)*	*Medium food grains demand (mt)*	*Demand water requirement (BCM)*	*High food grains demand (mt)*	*Demand water requirement (BCM)*
2010	249	489	265	536	271	556
2025	322	619	349	688	365	734
2050	469	830	539	1008	605	1191

Source: Hand Book of Agroculture 2009, ICAR, New Delhi

4.5.1 Traditional *vs* Improved Irrigation Methods

With increasing pressure on land and water, immediate action is needed to increase the irrigation efficiency to 60% for surface water and 75–80% for groundwater. The increased understanding of the soil-plant relationship has given rise to the concept that the best use of available water resources and optimum plant performance can be realized by prevention of moisture stress. Irrigation water may be applied to crops by flooding, by applying it under pressure or by applying it in drops or trickles depending on the source of supply of water, topography, quantity of water to be applied, the crop and method of cultivation of crop.

In surface methods of irrigation, water is applied directly to the soil surface from a channel or open ditch located at the upper reach of the field. The most common methods of surface irrigation are flooding, check-basin, basin, border-strip and furrow methods. Presently the irrigation efficiency of surface irrigation methods is low, but high efficiency can be obtained if water distribution systems are properly constructed to provide adequate control of water to the fields and proper land shaping is done for uniform distribution of water over the field. In surface irrigation, about 60–70% of the water released from the source is lost on the way by way of seepage, evaporation etc. as conveyance losses. Irrigation efficiency under different methods of irrigation (%) is shown in Table 4.24.

Table 4.24: Irrigation efficiency under different methods of irrigation (%)

Irrigation efficiency	*Methods of irrigation*		
	Surface	*Sprinkler*	*Drip*
Conveyance efficiency	40–50 (canal) 60–70 (well)	100	100
Application efficiency	60–70	70–80	90
Surface moisture evaporation	30–40	30–40	20–25
Overall efficiency	30–35	50–60	80–90

Source: Sivanappan (1988)

4.5.1.1 *Surface Irrigation Methods*

Flood irrigation

Flooding is the primitive uncontrolled surface irrigation method which is generally practiced in paddy field where standing water is required for long duration of crop growth and crop is cultivated in fairly level fields surrounded by dykes. This is the simplest method of irrigation and widely practiced in paddy dominated regions of the country. Under this system, the irrigation water is flooded on the entire surface of the field through a channel. In flooding method of irrigation, water is allowed from the channel into the field without much control on either side of the flow. The ideal size of each plot or basin is 0.1 to 0.2 ha for economizing water. The height of dykes around the basin should be 15–20 cm for impounding and effective use of rainfall. Large streams can be managed easily by this method. Uneven distribution and low water application efficiency are the common drawbacks of this method.

Check-basin

In check-basin method of irrigation, the field is divided into small plots surrounded by small bunds on all the four sides. Water from head channel is supplied to the field channels one after the other. Each field channel supplies water to two rows of check basins and water is applied to one basin after another. This method is adopted when the field is quite large and it is not easy to level the entire field. The size of the check-basins ranges from 4 m × 3 m to 6 m × 5 m depending on the stream size and soil texture. Bigger size streams of water need big check-basins. Light textured soils need small sized check basins. The advantages of this method is that water can be applied uniformly. Even with small streams (2 l/s), irrigation water can be applied efficiently. It is suitable for close growing crops like groundnut, wheat, finger millet, pearl millet. The main drawbacks of this method of irrigation is that more labour is required for field layout and irrigation. More land is wasted under channels and bunds and inter-cultivation is not possible due to obstruction by bunds.

Basin irrigation

Basin method is almost similar to check-basin method except that in later method, entire field is irrigated, while in basin method only the area around the trees are irrigated. Soil having medium to heavy texture *i.e.* loam to clayey soils with medium to low infiltration rates, low percolation losses and high water holding capacity are well suited for basin irrigation. Fruit crops or cucurbits vegetables are irrigated through this method. Basins are generally round in shape, occasionally square in shape. Basins may be connected by an irrigation channel. The basins are small when the trees are young and their size is increase with age of the trees.

Border irrigation

In this method, the water is supplied in small strips of land known as borders, which are uniformly graded. The bunds along with the borders help to guide the water to flow down the field. This method is commonly used to irrigate crops like wheat, pulses, maize and oil seeds. The borders can be dismantled after harvesting the crops. This irrigation method is suitable for medium to heavy textured soils. After delivering desired quantity of water to the border, the stream is turned off. Length of the strip ranges from 30 to 300 m and width from 3 to 15 m.

Table 4.25: Suggested size of borders and stream sizes for different soils and land slopes

Soil type	*Slope (%)*	*Width (m)*	*Length (m)*	*Flow (1itre/second)*
Sand	0.2–0.4	12–30	60–90	10–15
	0.4–0.6	9–12	60–90	8–10
	0.6–1.0	6–9	75	5–9
Loamy sand	0.2–0.4	12–30	75–150	7–10
	0.4–0.6	9–12	75–150	5–8
	0.6–1.0	6–9	75	3–6
Sandy loan	0.2–0.4	12–30	90–250	5–7
	0.4–0.6	6–12	90–180	4–6
	0.6–1.0	6	90	2–4
Clay loam	0.2–0.4	12–30	180–300	3–4
	0.4–0.6	6–12	90–180	2–3
	0.6–1.0	6	90	1–2
Clay	0.2–0.3	12–30	350	2–4

Source: Reddy and Reddy, (2009)

Compared to check-basins, this method is advantag-eous because labour required for field layout is less. Inter-cultivation is also possible as the strips are long. Irrigation streams can be efficiently used. However, distribution of water is not uniform due to more opportunity time for water at the strip leading to consuming more water.

Furrow irrigation

In this method the water is applied through a series of parallel furrows which are in turn connected with the main channel. Furrow irrigation is generally practiced for crops like sugarcane, potato, vegetables which are raised in ridges. The size and shape of the furrow depends on the crop grown and spacing adopted for the crop. Table 4.26 suggests length of furrows and stream sizes for different soils, land slopes and depth of water application.

Table 4.26: Suggested length of furrows and stream sizes for different soils, land slopes and depth of water application

Slope (%)	*Length of furrow (m)*								*Maximum flow (l/s)*
	Clay		*Clay loam and sandy loam*			*Loamy sand and sand*			
0.05	300	400	120	270	400	60	90	150	12
0.10	340	440	180	340	440	90	120	190	6
0.20	370	470	220	370	470	120	190	250	3
0.30	400	500	280	400	500	150	220	280	2
0.50	400	500	280	370	470	120	190	250	1.2
1.0	280	400	250	300	370	90	150	220	0.6
1.5	250	340	220	280	340	80	120	190	0.4
2.0	220	270	180	250	300	60	90	150	0.3
Application depth (mm)	75	150	50	100	150	50	75	100	

Source: Reddy and Reddy, (2009)

Sometimes the close-growing row crops like groundnut, linseed, etc., are occasionally given supplemental irrigation though they are originally planned as rainfed crops. Inter-cultivation is done so as to make shallow furrows. Applying irrigation through this shallow furrow is called corrugation. Vegetable crops like cucurbits are irrigated with short furrows of 5 to 6 in length and 5 to 6 rows are grouped into a basin. This method is useful for irrigation in kitchen gardens with small streams.

Other than normal furrow irrigation, some alternative methods of furrow irrigation have also been developed. In the normal method of furrow irrigation, water is allowed into every furrow at each irrigation which is called every-furrow irrigation. In every- furrow irrigation, water advances both laterally and downwards as water moves along the irrigated furrow and eventually the lateral wetted fronts from the adjacent furrows meet, Shortly, after this meeting, infiltration is predominantly downward. When irrigation water is not sufficient, water has to be saved without much reduction in crop yield. Some of the alternative method of furrow irrigation are paired row irrigation, skip irrigation, wide spaced furrow irrigation and within-row irrigation. General specifications for different surface methods for irrigation are given in Table 4.27.

Table 4.27: General specifications for surface methods of irrigation

Item	*Border*	*Check-basin*	*Furrow*
Soils	Most soils except coarse sands	Most soils	Most soils except coarse sandy
Infiltration rate	Moderately low tomoderately high	Moderate to low	Moderate to low
Slope (%)	0.20–1.0	No restriction	Minimum 0.05% upto 0.5% and even upto 2% in some cases
Stream size (l/s)	12–15	Large or small, basin, size to be adjusted	0.3 to 2.5 common maximum 12
Length (m)	60–300	Length: width 1.5–2.5:1	Usually 50–400, spacing 45–60 cm depending on the crop. Depth of furrow: 7.5–12.5 cm
Width (m)	6–30	3–	–
Crops	Wheat, barley, legumes, fodder, close growing crops	Groundnut, vegetables, rice, finger millet	Sugarcane, cotton, chillies, maize, vegetables

Source: Reddy and Reddy, (2009)

4.5.1.2 *Surge Irrigation*

In furrow irrigation, infiltration rate declines from initial high value at the beginning of irrigation to a low and stable value, generally at first quarter of the furrow. To reduce run-off, cut-back methods are used whereby the inflow is gradually reduced to match the changing soil infiltration rate. Surge irrigation provides opportunity to provide automation to this procedure. Surge irrigation is defined as the intermittent application of water to the field surface under gravity flow which results in a series of 'on' and 'off' modes of constant or variable time spans. Large intermittent flows rather than continuous ones are used with two sets of furrows and gated pipes laid in the 'Tee' configuration. Water is switched alternately from one set of furrows to the other by a valve and automatic time controller until irrigation is completed.

4.5.1.3 *Cablegation*

Another method of surface irrigation is Cablegation which is also an automated method of surface irrigation. The system can save labour and water compared to other surface application methods. Cablegation is a form of gated-pipe system. The gates or outlets are positioned near the top side of pipe and are always open. The spacing of the outlets corresponds to row spacing.

4.5.1.4 *Subsurface Irrigation*

In sub-surface irrigation, water is applied to the subsurface soil through underground perforated pipes. Water is also applied through deep trenches at 15 to 30 m intervals. Water gradually wets root zone through capillary movement. The surface soil is dry even when the root zone is wet and thus evaporation losses are minimized. Weed problem is less due to dry surface soil. The trenches are also used for drainage. This method is suitable where water table is shallow. However, maintenance of pipe line is a problem. Subsurface irrigation through trenches causes deep percolation losses.

4.5.1.5 *Micro-irrigation*

Micro-irrigation is an efficient method of providing irrigation water directly into soil at the root zone of plants. Thus, micro-irrigation minimizes such conventional losses as deep percolation, run-off and soil evaporation. It also permits the utilization of fertilizer, pesticides and other water-soluble chemicals along with irrigation water with better crop response.

The micro-irrigation system is becoming more and more popular in India, due to its advantages to produce more crop yield with better quality and saving in irrigation water. A properly designed micro-irrigation system can help growers of fruits, vegetables, flowers and other cash crops to save water, increasing yield; improving quality of produce, reducing labour cost reducing salt concentration in the root zone, allowing chemicals and fertilizers (soluble) to be used through the system, keeping inter-row spaces firm and dry and controlling diseases.

Among micro-irrigation, drip irrigation is more popular. It is the slow application of water in the form of discrete or continuous or tiny streams of miniature sprays through mechanical devices called emitters or applicators located at selected points along water delivery lines. Drip irrigation is known as trickle irrigation.

Though drip irrigation was introduced in India in the early seventies, significant development has taken place only in the eighties. The area under drip irrigation in India was 1,500 ha in 1985, increased to 6,000 ha in 1988 and was 24,500 ha in 1991. The growth of microirrigation has really gained momentum in recent years. This development has taken place mainly in areas of acute water scarcity in the states of Maharashtra, Andhra Pradesh, Karnataka, Tamilnadu and Gujarat. The important crops under micro-irrigation systems are coconut, grapes, banana, mango, chikoo, pomegranate, other fruit trees, plantation crop like sugarcane, cotton, groundnuts, vegetables, flower, etc

The Government of India through various agencies has sponsored many research schemes to the various institutions in the country from 1985 onwards. The results of various research studies undertaken under AICRP Plasticulture for the last few years are summarized in Table 4.28.

Advantages of drip irrigation with improved planting technique was demonstrated on Suru sugarcane in vertisol at Rahuri (Table 4.29). Among seven water saving techniques, paired row planting with drip irrigation was found the best for achieving higher water use efficiency (14.92 q/ha-cm).

4.5.1.6 *Sprinkler Irrigation*

Sprinkler irrigation system conveys water from the source through pipes under pressure to the field and distributes over the field in the form of spray of 'rain like' droplets. It is also known as overhead irrigation. Though sprinkler irrigation is known since 1946, it is only since 1980 that this system comes to be widely known to farmers in India. The introduction of light weight tubing and quick coupling systems made it a portable system and thereby economically competitive with surface methods of irrigation.

Sprinkler irrigation can be advantageously chosen in the following situations

- When the soil is too shallow eliminating the possibility of levelling of lands,

Table 4.28: Result of studies on micro-irrrigation

S. no.	*CROP*	*Yield (q/ha)*		*Irrigation (cm)*		*WUE (q/ha-cm)*		*Advantage of micro-irrigation (%)*	
		Surface	*Drip*	*Surface*	*Drip*	*Surface*	*Drip*	*Water saving*	*Increase in yield*
1	Beet	5.70	8.90	86.00	18.00	0.07	0.50	79.10	56.10
2	Bitter gourd	32.00	43.00	76.00	33.00	0.42	1.30	56.60	34.40
3	Brinjal	91.00	148.00	168.00	64.00	0.55	2.30	61.90	62.60
4	Broccoli	140.00	195.00	70.00	60.00	2.00	3.25	14.30	39.30
5	Cauliflower	171.00	274.00	27.00	18.00	6.30	15.20	33.30	60.20
6	Chilly	42.30	60.90	109.00	41.70	0.39	1.50	61.70	44.00
7	Cucumber	155.00	225.00	54.00	24.00	2.90	9.40	55.60	45.20
8	Lady's finger	100.00	113.10	53.50	8.60	1.87	13.20	84.00	13.10
9	Onion	284.00	342.00	52.00	26.00	5.50	13.20	50.00	20.40
10	Potato	172.00	291.00	60.00	27.50	2.90	10.60	54.20	69.20
11	Radish	10.50	11.90	46.00	11.00	0.23	1.10	76.10	13.30
12	Sweet potato	42.40	58.90	63.00	25.00	0.67	2.40	60.30	38.90
13	Tomato	61.80	88.70	49.80	10.70	1.24	8.28	78.50	43.50
14	Banana	575.00	875.00	176.00	97.00	3.27	9.00	45.00	52.20
15	Grapes	264.00	325.00	53.00	28.00	5.00	11.60	47.20	23.10
16	Papaya	130.00	230.00	228.00	73.00	0.60	3.20	67.90	76.90
17	Pomegranate	34.00	67.00	21.00	16.00	1.62	4.20	23.80	97.00
18	Water melon	82.10	504.00	72.00	25.00	5.90	20.20	65.30	51.30

Source: (Report of Task Force on Micro-irrigation, 2004)

Table 4.29: Yield and economics of suru sugarcane as influenced by planting methods and irrigation levels at Rahuri

Water saving treatments	*Yield of cane (t/ha)*	*Net monetary returns (Rs/ha)*	*B:C ratio*	*Water use efficiency (q/ha-cm)*
1. Paired row planting with surface irrigation	133.14	85570	2.40	9.65
2. Paired row planting with drip irrigation	144.04	96403	2.55	14.92
3. Normal planting with surface irrigation	118.14	68155	2.09	5.97
4. Normal planting with alternate furrow irrigation	103.53	56084	1.83	9.60
5. Four row planting with drip irrigation	122.60	73009	2.17	10.54
6. Four row planting with surface irrigation	128.60	81564	2.35	7.94
C.D at 5%	15.14	16116	0.27	–

Source: Annual Report, AICRP on Water Management, 2009–10

- When the land is too steep (>1%) for levelling economically,
- When light (<5 cm) and frequent irrigations are to be given,
- When soils are very sandy (rapidly permeable coarse textured soils), and
- When supplemental irrigation is to be given to dryland crops during prolonged dry spells, without any land preparation.

The saving in irrigation water (25–50%) by the use of sprinkler system is mainly due to conveying water through pipes from the source of irrigation to the field. The area that would be lost for cultivation under open ditches and bunds in surface irrigation system (approximately 5%) is saved when sprinkler system is used. In regions where light frost occurs occasionally, sprinkler irrigation is used to prevent the frost damage. Water application is more uniform with sprinkler system than with surface irrigation except during times of high winds.

As a case study, water requirement of wheat was fulfilled by sprinkler irrigation in clay soils of central India during 2005–08. Treatments consisted of four irrigation schedules by sprinkler *i.e.* IW/CPE, 0.6, 0.8, 1.0 and compared with surface irrigation at IW/CPE 1.0.

Results showed that different irrigation schedules significantly influenced wheat grain yield. Scheduling irrigation by sprinklers at IW/CPE 0.8 produced 3.96 t/ha wheat grain yield, significantly superior over IW/CPE 0.6 and surface irrigation at IW/CPE 1.0 (Table 4.30).

Table 4.30: Grain yield and water use of wheat under sprinkler irrigation during 2005–08 at Powarkhera (pooled yield of three years)

Treatment	*Yield (t/ha)*	*No. of irrigations*	*Total depth of irrigation (mm)*	*Water use efficiency (kg/ha-cm)*	*Water saving over surface method of irrigation*	
					Depth (mm)	*Percent*
Sprinkler irrigation at IW/CPE 0.6	3.4	2	150	231.2	200	57
Sprinkler irrigation at IW/CPE 0.8	3.96	3	200	197.8	150	43
Sprinkler irrigation at IW/CPE 1.0	4.2	4	250	170	100	29
Surface irrigation at IW/CPE 1.0	3.9	4	350	112.3	–	–
CD (5%)	0.2	–	–	–	–	–

Source: Annual Report, AICRP on Water Management, 2009-10

Water requirement of maize through sprinkler irrigation in relation to mulching was studied in Vertisols of Malaprabha command area of Belavati during 2008 (Table 4.31). The results indicated that sprinkler irrigation level at 0.8 IW/CPE produced higher maize grain yield of 82.86 q/ha, gross income of Rs. 53,860/ ha, net income of Rs. 35,610/ ha and B:C ratio of 2.96 but all these are at par with the results obtained in sprinkler irrigation level 0.6 IW/CPE. A significantly higher water use efficiency of 18.05 kg/ha mm was obtained in sprinkler irrigation level 0.6 IW/CPE. The water saving in sprinkler irrigation level 0.6 IW/CPE was 12.0 per cent over 0.8 IW/CPE.

4.5.2 Concept of Irrigation Performance Indicators

4.5.2.1 *Irrigation efficiencies*

One of the major communication barriers within the irrigation and water management debate is the inappropriate use of terms and data. A large

Table 4.31: Yield and economics of maize under sprinkler and surface irrigation methods in Belavati

Treatment	*Maize grain yield (q/ha)*	*WUE (Kg/ha -mm)*	*Gross income (Rs.)*	*Net income (Rs.)*	*B:C ratio*
Irrigation levels					
I_1 = 0.8 IW/CPE (Sprinkler)	82.86	16.57	53860	35610	2.96
I_1 = 0.6 IW/CPE (Sprinkler)	79.41	18.05	51618	33368	2.83
I_1 = 0.4 IW/CPE (Sprinkler)	59.43	16.98	38626	20376	2.12
I_1 = 0.6 IW/CPE (Surface)	70.45	16.01	45793	27543	2.51
CD (0.05)	3.67	0.83	2387	2387	0.14
Mulches					
M_1 = Maize straw @ 5t/ha	77.01	17.82	50096	30596	2.57
M_2 = Without mulch	69.00	15.98	44853	27853	2.64
CD (5%)	2.26	0.52	1471	1471	0.08

Source: Annual Report, AICRP on Water Management, 2009–10.

number of indices for the assessment of irrigation performance have been proposed. It is difficult to adequately evaluate irrigation performance using a single parameter. At the system or whole farm level, a range of performance parameters may be appropriate depending on the spatial and temporal boundary conditions established for the evaluation. Similarly, at the field scale, the adequacy of the irrigation is dependent on the water stored within the crop root zone, the water percolating below the roots zone, surface runoff, the uniformity of the applied water distribution, and the unused storage capacity of the soil profile following irrigation. Some terms reflect the performance only of the agronomic component of the system, while others focus on the engineering component. Data on system performance may include either all of the sources of water to the crop (*e.g.* irrigation + rainfall + soil water) or only the irrigation component (either delivered to the farm gate or applied to the field). In all cases, it is important that the terms used are well defined.

Starting from the storage to the point of use in agricultural fields, the movement of water in a surface water based irrigation system can be viewed as four separate operations *i.e.*, (i) diversion from storage system, (ii) conveyance in main and branch or lateral or secondary canals, (iii) distribution and conveyance by distributary or tertiary canals, and (iv) field application by minor or quaternary or farm canals either directly or through field channels. The applied water is taken up by the plant roots to meet its consumptive use demand, and hence it may be considered as a separate operation.

Therefore the following water management sub-systems may exist on irrigated farms:

- Supply systems (*e.g.* harvesting or lifting from canal and captured overland flows; pumping groundwater from well; and/or supply from irrigation dams, channels and/or pipes)
- On-farm storage systems
- On-farm distribution systems (*e.g.* earthen channels; gated pipes; or pressurized enclosed systems)
- Application systems (*e.g.* surface, spray, micro-systems) and recycling systems

Efficiency is generally defined as a dimensionless ratio of output over input. Israelsen (932) was probably the first to define irrigation efficiency as: "the ratio of irrigation water transpired by the crops of an irrigation farm or project during their growth period, over the water diverted from a river or other natural source into the farm or project canal or canals during the same period of time". Since then, different types of irrigation efficiencies have been expounded.

The efficiency of water use can be defined for each of the above sub-systems based on the volumetric water inputs and outputs, or uses and losses (Table 4.32).

Table 4.32: Recommended irrigation efficiency definitions

Term	*Definition*
Overall Project Efficiency (*Ep*)	$\dfrac{\textit{Irrigation water available to crop}}{\textit{Total inflow into system supply}}$
Conveyance Efficiency (*Ec*)	$\dfrac{\textit{Total outflow from system supply}}{\textit{Total inflow into system supply}}$
Distribution Efficiency (Ed)	$\dfrac{\textit{Water received at field inlets}}{\textit{Total outflow from system supply}}$
Field Application Efficiency (*Ea*)	$\dfrac{\textit{Irrigation water available to crop}}{\textit{Water received at field inlet}}$

Reservoir storage efficiency (E_s)

Jensen *et al.* (1967) defined the reservoir storage efficiency as:

$$E_s = \frac{\text{Volume of water delivered from the reservoir}}{\text{Volume of water delivered to the reservoir, surface or ground water, for irrigation}}$$

E_s measures the fraction of storage utilized for irrigation. It also measures the volume of water lost or retained in the reservoir.

In India, several diversion storage structures have been constructed across rivers and streams to divert runoff the rivers and the streams to either canals or field channels for the purpose of irrigation. Such structures detain small volumes of live storage, but over a period of time divert large quantity of water for irrigation and other usage. Whenever, water is not needed for irrigation, it is allowed to flow in the bed of the river/stream. In order to depict such type of situation, Singh (1996) defined the reservoir detention factor, E_d as:

$$E_d = \frac{\text{Volume of water delivered to the reservoir for irrigation and other useful purpose}}{\text{Live storage of the reservoir}}$$

Minor irrigation reservoirs in medium to high rainfall zones may get filled several times during a cropping season/year. The reservoir detention factor defined by above equation is a measure of the number of live fillings of the reservoir utilized beneficially.

A review of 90 irrigation projects through out world indicated generally low irrigation efficiencies, with only 20–40% of the water diverted from the reservoir being effectively used by the crop, while in India, the irrigation efficiency is around 10–20%. The main components of water losses in an irrigation system are: (a) water losses in storage (10 to 20%) (b) water losses in conveyance systems (25 to 40%) (c) water losses in operation (10 to 30%) and (d) water losses in application (45 to 70%). Irrigation efficiency is considered to be as low as 10–20% in some projects in India. It is estimated that 11 million ha-m of water is contributed to ground water recharge annually through seepage from canal systems and the return flow from irrigation.

In no irrigation project in India, the total losses in the canal distribution system and field has been less than 50% of the head discharge. Studies conducted in Nagarjunasagar Project have revealed that only 40–60% of water released from the reservoir reaches the field and only 20–40% is used by the crops (Reddy and Reddy, 2009).

An overall irrigation efficiency of 40–50% for a project can be considered to be high and this can be achieved only by the adoption of improved irrigation and drainage practices based on scientific land and water management principles. Raising irrigation efficiency has following benefits:

- A larger area can be irrigated with the same volume of water.
- A water shortage can be made less severe by increasing the relative water supply value.
- The waterlogging and salinity can be reduced or eliminated thereby saving investment on drainage.
- There is a reduction in competition with other water users.
- There is a saving in energy due to reduction in pumping for irrigation and drainage.

As per the National Commission on Integrated Water Resources Development Plan (NCIWRDP), 35 to 40 per cent efficiency in surface water and 65–70 percent efficiency in ground water will be a fair approximation.

A major concern with the sole use of volumetric efficiency terms (Table 11) for irrigation evaluation is that they do not provide any assessment of the overall irrigation performance in relation to crop production and economic returns. Terms such as "crop water use efficiency" and "marginal crop water use efficiency", which include both yield and volumetric components are more relevant to evaluate irrigation performance (Table 4.33).

Factors influencing irrigation efficiency

The main objective of irrigation management is to supply a desired amount of water to a crop at a specified time. Hence, effective irrigation management requires some knowledge about how much water to apply (the irrigation volume) and when to apply the water (the irrigation schedule). Adequate prediction of the irrigation volume and schedule requires for both the on- and off-farm water management and also to have some knowledge of the water availability within the crop root zone and the rate of crop water use.

Table 4.33: Water use efficiency indices for the evaluation of irrigation performance

Term	*Definition*
Crop Water Use Index	$\frac{\text{Production (kg)}}{\text{Evapotranspiration (mm)}}$
Gross Production Water Use Index	$\frac{\text{Total Product (kg)}}{\text{Total Water Applied }(m^3)}$
Irrigation Water Use Index	$\frac{\text{Total Product (kg)}}{\text{Irrigation Water Applied }(m^3)}$
Marginal Irrigation Water Use Index	$\frac{\text{Marginal Production due to irrigation (kg)}}{\text{Irrigation Water Applied }(m^3)}$
Gross Production Economic Water Use Index	$\frac{\text{Economic return (Rs)}}{\text{Total Water Applied }(m^3)}$
Irrigation Economic Water Use Index	$\frac{\text{Economic return (Rs)}}{\text{Total Irrigation Water Applied }(m^3)}$
Marginal Irrigation Economic Water Use Index	$\frac{\text{Marginal return due to irrigation (Rs)}}{\text{Irrigation Water Applied }(m^3)}$
Crop Economic Water Use Index	$\frac{\text{Economic Return (Rs)}}{\text{Evapotranspiration (mm)}}$

Irrigation design and management decisions are the result of a complex interaction of many variables which are rarely consistent between individuals. Irrigation management is also often expected to maximize efficiencies and minimize the labour and capital requirements of the particular irrigation system without adversely affecting the growing environment for the plant. Supply volume and rate constraints imposed at the system level may significantly affect the performance and efficiency at the farm and field scales. Hence, irrigation design and management practices are influenced by a wide range of factors including:

- Agronomic (*e.g.* crop responses to climatic and soil moisture variables),
- Environmental (*e.g.* climate, soils, topography),
- Social (*e.g.* experience, education, labour availability),
- Economic (*e.g.* capital availability, operating costs, returns from product),
- Historical (*e.g.* existing infrastructure, previous farming systems),

- Hydrological (*e.g.* river flow regimes, groundwater issues; surface flow harvesting),
- Engineering constraints (*e.g.* hydraulic design limitations on pumps, pipes and storages, supply capacities) and
- Regulatory policy and administrative procedures

Many irrigation management actions are dependent on the specific type of irrigation application system or design available. Other decisions (*e.g.* frequency of irrigation, depth of water to be applied) are common to all systems and dependent on the nature of the crop, soil and environmental conditions.

Factors influencing the efficiency of storage, distribution and recycling Systems

The operational efficiency of storage, distribution and recycling structures is influenced by both design and management variables as well as environmental factors. In particular, consideration should be given to:

- Evaporative surface area/volume ratio,
- Wind velocity/surface protection,
- Seepage/drainage rate (function of construction material and head) and
- Period of storage/distribution

Factors influencing the efficiency of application systems

The ability of the in-field irrigation system to apply water efficiently and uniformly to the irrigated area is a major factor influencing the agronomic and economic viability of the production system. The performance evaluation of in-field application systems can be divided into the two major components of water losses and uniformity of application. Although both components are influenced by system design and management practices, the losses are predominantly a function of management while the uniformity is predomi-nantly a function of the system design characteristics. However, the irrigation system is not usually expected to supply all of the moisture required for crop production as some of the crop's water requirements may be met by preseason moisture stored in the soil profile, rainfall during the growing season, or from shallow groundwater tables. Hence, optimal irrigation management requires not only knowledge of the characteristics of the application system but an understanding of the environment in which it operates.

The major sources of water loss by in-field application systems are due to evaporation (from the atmosphere, free water surface or soil surface), deep drainage or by surface run-off. The dominant loss mechanism is closely related to the method of application, but in all cases may be substantially reduced by the adoption of appropriate management practices. Higher application efficiencies can normally be expected through the use of micro-irrigation or low pressure overhead spray systems. However, substantial water losses are often found where these systems are being used with inappropriate management practices (*e.g.* excessive watering periods, irrigating in high wind).

Other influences on the efficiency of the application system include:

- Infiltration characteristics (soil, wetted area, water quality, compaction),
- Application technique,
- Field layout,
- Application rate and period (continuous, surge/pulse, cut-back) and
- Application spatial pattern (bed size, row spacing, emitter spacing).

Factors influencing crop water use

Crops have a variable capacity to extract water from the soil and to convert this water into plant products. Hence, the efficiency of crop water use varies considerably between different crop species. However, there are also considerable differences in crop water use between different varieties (or cultivars) of the same species. Other factors affecting the crop water use include cultural practices (*e.g.* planting date, spacing, cultivation practice, mulching), irrigation management (*e.g.* irrigation interval, prior plant stress, water placement) and environmental influences (*e.g.* nutrition, radiation, temperature, drainage).

Irrigation design and management options to improve water use efficiency

The strategies to improve water use efficiency revolve around the central themes of reducing losses out of the system (*i.e.* evaporation, deep drainage, and run-off), reducing crop evapotranspiration during non-critical periods, and increasing the effectiveness of stored soil moisture and rainfall during the season. The efficiency of storage and distribution systems is influenced most significantly by the system design and construction techniques (influences evaporation/seepage/failure). Water savings at the field scale may be achieved by:

- Maximizing the pre-season soil moisture storage
- Minimizing evaporation losses
- Minimizing crop transpiration while maintaining agronomic and economic goals
- Maximizing net effective precipitation during the growing season
- Improving the application efficiency of the irrigation application system and
- Reducing deep percolation to only that necessary for leaching.

Uniformity coefficient

The field application efficiency gives no information about the uniformity of applied water in the field. The uniformity coefficient (C_u) introduced by Christiansen (1942), or some other statistical measure of uniformity can be used to quantify the uniformity of water application. The C_u, defined by Christiansen is as below;

$$C_u = \frac{1.\,(\frac{1}{n}\sum_{i=1}^{n}(D_i - D_f)}{D_f}$$

where,

D_i = Depth of water applied to the field at the i_{th} point of observation

D_f = Mean depth of irrigation water furnished to the field

N = Number of observations

The quantity in the parentheses gives the average absolute deviation of the observations from the mean. Sometimes C_u is expressed in percentage by multiplying its fractional value by 100.

Bos (1980) recommended the use of the following ratios (irrigation efficiencies) to describe efficiency of water use at crop production level:

Yield: ET ratio (R_{ye}):

$$R_{ye} = Y_i - Y_r/ET_i - ET_r$$

where,

Y_i = Mass of marketable crop produce with irrigation

Y_r = Mass of marketable crop that could be produced with irrigation.

ET_i = Mass of water evapotranspired by the irrigated crop

ET_r = Mass of water that would be evapotranspired by the same crop if not irrigated.

Yield: Water Supply Ratio (R_{yw}):

$$R_{yw} = Y_i - Y_r/W_f$$

where,

W_f is mass of irrigation water furnished to the field.

Yield: Water Cost Ratio (R_{ye}):

$$R_{yc} = C_i - C_r/C_f$$

where,

Ci = Value of crop produced with irrigation

Cr = Value of crop that could be produced without irrigation

Cf = Cost of applied irrigation water including amortization, labour and supplementary production costs

Yield: Water Expense Ratio (R_{ywe}):

$$R_{ywe} = Y_i/W_e$$

Where, W_e = total water expense in obtaining the yield. W_e is the sum of water supplied from irrigation, rainfall, soil profile and capillary rise. Thus, it considers total water spent in obtaining the crop yield. Higher R_{ywe} would indicate efficient utilization of water. The R_{ywe} of selected crops indicates that higher yield does not necessarily mean higher water expense efficiency.

4.5.3.2 *Concepts of water productivity*

Concepts of water productivity in agricultural production systems is focused on producing more food with the same water resources or producing the same amount of food with less water resources. Water productivity is the amount of crop production per unit amount of water used. Earlier irrigation efficiency or water use efficiency was used to describe the performance of irrigation systems for producing more yields. Now water productivity is used to measure the amount of money earned per unit quantity of water. But definition of water productivity changes with the background of the researcher or stakeholder involved. For example, obtaining more kilograms dry matter production per unit of transpiration is a key issue for plant breeders. At a basin scale, economists wish to maximize the economical value from water used. Water managers tend to be more concerned with the total water input. Rainfed farmers in arid areas are highly concerned with doing the most with the limited rainfall. For irrigation farmers and managers factors and they will evaluate their own water productivity on the basis of canal water supplies in relation to crop yield, rainfall, supplemental irrigation or full irrigation supplies.

Table 4.34: Stakeholders and definitions in the water productivity framework

Stakeholders	*Definitions*	*Scale*	*Target*
Plant physiologist	Dry matter/transpiration	Plant	Utilize light and water resources
Nutritionist	Calorie/transpiration	Field	Healthy food
Agronomist	Yield/evapo-transpiration	Field	Sufficient food
Farmer	Yield/supply	Field	Maximize income
Irrigation engineer	Yield/irrigation supply	Irrigation scheme	Proper water allocation
Groundwater policy maker	Rupees/ground water extraction	Aquifer	Sustainable extraction
Basin policy maker	Rupees/evapo-transpiration	River Basin	Maximize profits

4.5.2.2 *Irrigation scheduling*

Scheduling of irrigation means when and how much water are be applied to the crops. Objectives of scheduling irrigation are to obtain higher crop yields and maximum water-use efficiency with least damage to soil productivity. Irrigation scheduling decisions are to be made for mainly two situations: (i) when adequate water is available on demand, and (ii) when only limited water is available. In both the situations, scientific irrigation schedules should be based on soil-water-plant-atmosphere relationship.

Irrigation scheduling is done basically by three approaches. (a) meteorological approach, (b) soil indicator approach (c) plant indicator approach. Out of these three, meteorological approach and soil indicator approach are more common.

A brief description of approaches for irrigation scheduling is presented below.

Transpiration approach

Transpiration ratio is the quantity of water required to produce a unit quantity of dry matter. Some considered dry matter to be the weight of the shoot, while others considered it as the weight of the economic produce of the crop. This ratio was considered to be the water requirement of the crop, but later this idea was abandoned because under field conditions, dry matter production and water-use are affected by different sets of other factors.

Soil-water content approach

Irrigation scheduling on the basis of soil water content consists of two assumptions: (i) assuming threshold moisture content below which the plant growth is adversely affected, or (ii) determination of optimum fixed time interval between successive irrigations. The criterion of soil moisture content was soon discarded as the assumptions, threshold moisture content as well as fixed time interval, vary with soil texture.

Available soil moisture depletion approach

This approach essentially consists of determining the lower limit of available water in an assumed shallow root zone of the crops at which irrigation must be applied to avoid yield reduction. This approach has been widely used and available soil moisture depletion percentage has been fixed for many crops.

Soil-moisture tension approach

Soil-moisture tension has been used as an index of scheduling irrigation using tensiometers and gypsum blocks installed at specific soil depths to monitor soil moisture tension. Mean soil-moisture tension integrated across entire root zone is a more reliable irrigation guide than soil-moisture tension of a shallow soil layer. The approach is more useful for orchards and vegetable crops, particularly on coarse textured soils where most of the available water is held at low tensions. In soils, capable of withstanding higher soil-moisture tension, properly calibrated gypsum blocks can be used as a tool for irrigation scheduling.

The use of tensiometers and gypsum blocks could not become popular due to many limitations. The tensiometers do not work beyond 0–0.8 bar range, Gypsum blocks do not work at very low moisture tension and also fail to establish close contact in some swelling-shrinking type soils. Further, these approaches do not indicate the amount of irrigation water applied directly but require the use of soil-moisture characteristic curve to interpret the soil moisture content. There is a considerable time lag for the porous sensing parts of the tensiometers and gypsum blocks, to equilibrate with the surrounding soil and as such the soil moisture tensions do not give instantaneous values.

Meteorological approach

Meteorological approach of scheduling irrigation is also based on two assumptions: (i) evapo-transpiration (ET) of the crop is generally a fraction or at the most equal to evaporation from free water surface, and (ii) the ET at potential rate is considered desirable for optimum growth and yield of a crop. Irrigation can be scheduled if allowable depletion of the water in root zone and ET for short periods during the growing seasons is known. Periodic ET of a crop can be experimentally determined with the use of field lysimeters. At the end of each such period, the crop sown after charging the soil profile to field capacity would require an irrigation of an amount sufficient to meet the sum total of cumulative ET minus rainfall during the period since previous irrigation. Prihar *et al.* (1974) translated this approach to a more practical approach called IW/CPE approach. IW/CPE is the ratio between a fixed amount of irrigation water (IW) and cumulative open pan evaporation (CPE) minus effective rainfall since previous irrigation. The approach has been evaluated extensively to develop optimum irrigation schedules for different crops and agro-climatic regions of the country. The approach has also been compared with irrigation scheduling based on the growth stages of the crop and found to consume less water.

Some other methods of irrigation scheduling are based on plant indicators approach, Visual plant symptoms, Plant water content and water potential, Leaf diffusion resistance (LDR), plant temperature based approach.

Critical growth stages of different crops

As a result of extensive experimentation, critical growth stages of various crops have been identified in respect of water demand (Table 4.35). Irrigation needs to be applied at critical growth stages for maximum water use efficiency.

Table 4.35: Critical growth stages of some important crops in respect of water demand

Crop	*Critical crop growth stages*
Wheat	CRI, flowering
Rice	Transplanting to tiller initiation, panicle initiation to flowering
Maize	4–5 leaves, knee-high, taselling, silking, grain formation
Sorghum	Booting , vegetative
Pearl millet	Flowering
Pigeon pea	Flower initiation, pod development
Chick pea	Branching, pod development
Kidney bean	Vegetative, flowering, pod development
Green gram (Summer)	Vegetative, flowering, pod development
Soyabean	Flowering and pod development
Groundnut (Kharif)	Pegging and pod development
Groundnut (Rabi)	Vegetative, branching, flowering, pegging & pod development
Mustard	Flowering
Sunflower	Vegetative, disc formation, flowering
Sesame	Flowering

Source: Yadav *et al.* (2000)

In Orissa condition, based on critical growth stages, irrigation was given to grow 5 winter crops and water use efficiency of crops under irrigated and rainfed conditions were compared by Kar *et al.* (2005) (Table 4.36).

Table 4.36: Water use efficiency (kg ha^{-1}mm^{-1}) by different crops (rainfed *vs.* irrigated) based on irrigation scheduling at different critical growth stages

Crops	*Irrigation based on phenology*	*WUE (kg ha^{-1} mm^{-1})*		
		Y1	*Y2*	*M*
Linseed	(i) Vegetative +50% flowering+grain filling	2.18	2.32	2.25
	(ii) Vegetative +50% flowering	2.27	2.44	2.35
	(iii) 50% flowering	1.54	1.51	1.52
	(iv) Rainfed	0.67	0.67	0.74
	S.E.m (±)	0.37	0.41	0.37
Safflower	(i) Vegetative +50% flowering+grain filling	2.58	3.08	2.83
	(ii) Vegetative +50% flowering	1.96	2.29	2.12
	(iii) 50% flowering	1.45	1.67	1.56
	(iv) Rainfed	1.22	1.26	1.24
	S.E.m (±)	0.31	0.39	0.35
Gram	(i) Vegetative +50% flowering+grain filling	1.61	1.83	1.72
	(ii) Vegetative +50% flowering	1.34	1.47	1.40
	(iii) 50% flowering	0.73	1.02	0.87
	(iv) Rainfed	0.33	0.40	0.37
	S.E.m (±)	0.29	0.31	0.29
Pea	(i) Vegetative +50% flowering+grain filling	3.11	3.55	3.33
	(ii) Vegetative +50% flowering	3.18	3.80	3.49
	(iii) 50% flowering	1.66	1.93	1.79
	(iv) Rainfed	1.35	1.44	1.38
	S.E.m (±)	0.48	0.59	0.53
Mustard	(i) Vegetative +50% flowering+grain filling	2.13	2.27	2.20
	(ii) Vegetative +50% flowering	1.44	1.66	1.55
	(iii) 50% flowering	1.01	1.11	1.06
	(iv) Rainfed	0.53	0.45	0.49
	S.E.m (±)	0.34	0.39	0.36

Source: Kar *et al.* (2007b)

Results of an experiment of Kota, Rajasthan on water expenses efficiency of wheat under five irrigation schedules (6 cm irrigation at CRI, two irrigation at CRI + milking stage, at 40% depletion of available soil moisture, at 60% depletion of available soil moisture and four irrigations at critical growth stage) with four wheat cultivars (HI-8498, Raj-3765, GW-173 & Raj-3077) are given in Table 4.37. Results showed that maximum average wheat yield of 4.97 t/ha was obtained when

Table 4.37: Effect of irrigation schedules on the grown yield and water expense of different wheat varieties during 2005–08 at Kota

Treatments	*Wheat yield (t/ha)*					*WEE (kg/ha-cm)*			
	2005	*2006*	*2007*	*2008*	*Mean*	*2005*	*2006*	*2007*	*2008*
Irrigation schedule									
One irrigation at CRI	2.36	2.38	3.92	2.58	2.81	147.37	148.44	206.21	161.78
Two irrigations at CRI & milk stage	3.37	3.98	4.17	3.69	3.81	153.36	180.82	220.00	167.82
Irrigation at 40% depletion of ASM	4.48	4.99	5.11	5.31	4.97	157.0	175.00	179.12	189.71
Irrigation at 60% depletion of ASM	3.75	4.16	4.82	4.78	4.37	120.80	134.03	155.32	154.35
Four irrigation at critical stages	4.30	4.85	4.97	4.83	4.74	126.32	142.50	146.15	142.21
$CD_{(0.05)}$	0.43	0.59	0.56	0.37	–	–	–	–	–
Varieties									
HI-8498	3.92	4.35	5.01	4.33	4.40	137.44	152.49	175.61	154.64
Raj-3765	3.79	4.23	4.49	4.01	4.13	132.91	148.24	157.51	143.43
GW-173	3.27	3.89	4.81	4.42	4.09	114.56	136.35	168.67	158.11
Raj-3077	3.50	4.30	4.09	4.19	4.02	122.70	150.70	143.33	149.64
$CD_{(0.05)}$	0.27	0.22	0.35	0.27	–	–	–	–	–

Source: Annual Report, AICRP on Water Management, 2009–10

irrigations were applied at 40% depletion of available soil moisture (ASM). The yield at this treatment was statistically at par with that obtained in four irrigations at critical growth stages. The maximum water expanse efficiency was recorded under irrigation at 40% depletion followed by two irrigations at CRI & milk stage, one irrigation at CRI, irrigation at 60% depletion and four irrigations at critical stages.

4.5.2.3 *Water and irrigation requirements of different crops*

Water requirement is the quantity of water required by a crop in a given period of time for its normal growth under field conditions at a particular place. This includes evapo-transpiration or consumptive use, application losses and water requirement for special operations such as land preparation, transplanting, puddling, leaching etc. The irrigation water requirement refers to the total water requirement of crops exclusive of effective rainfall and contribution from soil profile.

The water requirement of the crops is determined experimentally in weighing type lysimeters from a water-balance taking into consideration the water added, water retained in soil, and water losses through all sources, namely, evaporation, transpiration and deep percolation. Though weighing type lysimeters provide precise and direct measurement of amount of water applied and lost by the crops, these do not simulate field conditions such as temperature, water-table, soil-texture and density, representative of field conditions. The cost of lysimeters and their installation prohibit use for determining water requirements at large number of places and for a large number of crops.

The crop water requirements are, therefore, determined from experimental plot studies from measurement of water supplies to the field and changes in soil moisture on a seasonal basis. The seasonal water requirements are computed by adding measured quantities of water at predetermined intervals, effective rainfall during the period, and profile water contribution as given by equation below.

$$WR = IR + ER + \frac{M_{bi} - M_{ei}}{100} / 100 \times A_i \times D_i$$

when,

WR= Seasonal water requirement, mm

IR = Total irrigation water applied, mm

M_{bi} = Moisture (%) at the beginning of the season in the ith layer of the soil

M_{ei} = Moisture (%) at the end of the season in the ith layer of the soil

ER = Effective rainfall, mm

A_i = Apparent specific gravity of ith layer of the soil, g cm^{-3}

D_i = Depth of the ith layer of soil within the root zone, mm,

and

N = Number of soil layers in the root zone.

The net irrigation requirement is the depth of irrigation water exclusive of precipitation, carry-over soil moisture or ground water contribution or other gain in the soil-moisture that is required for crop production. It is the amount of irrigation water required to bring the soil moisture level to field capacity in the effective root zone. An important consideration in water-balance studies is the period or time interval over which the balance is made. Too short a period might be impractical; while too long a period might mask the occurrence of short-term critical stages. It is necessary to ensure that the crop-water-balance is maintained positive continuously during the growing season.

A considerable work has been done on irrigation scheduling on different crops like rice, wheat, sugarcane, cotton, millets, pulses, oilseeds and other crops by Directorate of Water Management, Bhubaneswar and AICRP on Water Management. The recommended irrigation requirement and number of irrigations of different field crops is given in Table 4.38.

But, it is difficult to generalize the recommendations in such a vast country like India. Irrigation schedules have been largely developed in experimental fields, but are rarely used precisely in farmer's field. These scientific approaches require some measurement at the field level and are difficult for farmers to adopt because of their small holdings and uncertainties in availability of water. The results of an on-farm experiment conducted at Orissa on water use and water use efficiency of 5 winter crops are given in Table 4.39 below.

A comparison of all these irrigation scheduling approaches leads to the conclusion that an estimation of permissible soil moisture

Table 4.38: Recommended irrigation requirement and number of irrigations for different crops

Crop	*Irrigation requirement (mm)*	*Number of irrigations*
Cereals and millets		
Wheat	300–400	5–6
Rice	750–1640	10–20
Sorghum	140–300	1–4
Maize: rainy season	100–350	2–5
Summer season	500–900	10–18
Pearl millet	150–250	2–3
Barley	100–300	2–5
Pulses and oilseeds		
Chickpea	80–160	1–4
Pigeonpea	70–140	1–2
Groundnut	150–750	3–10
Mustard	60–180	1–3

Source: Yadav *et al*., (2000)

depletion is essential before the crop yield starts declining. Among the approaches, the meteorological approach (IW/CPE) is widely used because of its simplicity when the water supply is adequate. When water is inadequate, the irrigation may be scheduled at critical growth stages or at variable IW/ CPE ratios. However, emphasis has to be shifted from the potential yield to the optimum yield with maximum water use efficiency. The optimum depth of irrigation water and crop water use and irrigation requirements were standardized on various crops in different agro-climatic regions of the country through AICRP on Water Management and are given in Tables 4.40, 4.41 and 4.42.

4.5.3 Associated Problems of Irrigation Development

Faulty management of surface and ground water irrigation resources has led to serious implications on soil health and on crop production and posed threat to agricultural sustainability of irrigated ecosystems. Indiscriminate use of water has created problems like soil salinity and water logging resulting in productivity decline. Important problems have emerged due to non-judicious use of irrigation water are discussed below.

4.5.3.1 *Rise in water table*

In many canal command areas, unwise management of irrigation water has resulted in rising of water table and the high water table condition

Table 4.39: Yield and water use efficiency (pooled data of two years) of different crops with different levels of irrigation

Crop	*Grain yield (kg ha^{-1})*				*Water use (mm)*				*Water use efficiency (kg ha^{-1} mm^{-1})*			
	I_2	I_3	I_4	S.E (±)	I_2	I_3	I_4	S.E (±)	I_2	I_3	I_4	S.E (±)
Maize	1845	2950	5805	1179.7	334	385	438	30.0	4.95	7.25	13.8	2.58
Groundnut	785	1020	1590	207.2	348	395	444	27.7	2.24	2.58	3.35	0.32
Sunflower	905	1205	1715	205.7	395	429	476	23.4	2.29	2.60	3.39	0.31
Wheat	1420	2250	2780	1270.2	348	390	440	28.8	4.08	6.31	5.85	0.57
Potato	8050	9650	12400		355	394	432	22.2	22.7	24.5	28.7	1.77

I_2 = Two irrigations, I_3 = Three irrigations, I_4 = Four irrigations. *Source:* Kar *et al.*, (2005)

Table 4.40: Grain yield, irrigation water requirement, and water use in major coarse cereals at optimum schedule of irrigations (based on IW/ CPE) at different locations

Location	*Soil type*	*No. of irrigation (depth, cm)*	*Optimum IW/CPE*	*Grain yield (t/ha)*	*Irrigation water requirement (cm)*	*Water use (cm)*
Winter maize						
Pusa (Bihar)	Sandy loam	5(6)	1	3.81	30	33.3
Madhepur (Bihar)	Sandy loam	4(6)	1	3.79	24	39
Rahuri (M.S)	Clay loam	2–3(8)	0	3.53	20	37.8
Faizabad (U.P)	Silt clay loam	5(6)	0.60	5.76	30	35.3
Morena (M.P)	Clay loam	8–9(7.5)	0.90/1.20	5.22	64	70.1
Hyderabad (A.P)	Sandy loam	8–9(5)	0.90	3.70	43	46
Dharwad (Karnataka)	Black clay	6–7(6)	0.80	5.01	39	47.5
Coimbatore (T.N)	Sandy clay loam	6(6)	0.60	3.55	36	40
Kota (Rajasthan) Sorghum	Clay loam	2–3(6)	0.80	2.67	15	54.3

Table 4.40: (*Contd...*)

Table 4.40: (*Contd...*)

Location	*Soil type*	*No. of irrigation (depth, cm)*	*Optimum IW/CPE*	*Grain yield (t/ha)*	*Irrigation water requirement (cm)*	*Water use (cm)*
Kota (Rajasthan), *Kharif*	Clay	3(6)	0.75/0.80	3.14	18	56
Navasari (Gujarat), *Summer*	Black clay	6(8)	0.90	5.13	48	60*
Parbhani (M.S),	Black clay	1(6)	0.40	4.02	624	45.8*
Kharif	Black clay	4(6)	0.60	3.44		24*
Rabi						
Dharwad (Karnataka), *Rabi*	Black clay	4(6)	0.40	4.03	24	44.9*
Bhavanisagar (T.N), *Kharif*	Sandy loam	5(5)	0.40	3.01	25	51.4*
Rahuri (M.S), *Rabi*	Black clay	5(7)	1.05	3.73	35	40.8*
Pearl millet						
Sriganganagar (Raj.), *Kharif*	Sandy loam	3(6)	0.55/0.60	1.94	18	35.5*
Kota (Rajasthan), *Kharif*	Sandy loam	3(6)	0.75/0.80	2.84	18	58.3*
Rahuri (M.S), *Rabi*	Black clay	4(5)	0.55/0.60	2.40	20	34*
Hyderabad (A.P.), *Rabi*	Sandy loam	10(5)	0.75/0.80	2.20	50	53*

Source: Annual Reports of All India Coordinated Research Project on Water Management for 1972–73 to 1996–97

Table 4.41: Crop yield, irrigation water requirement, and water-use in cotton and sugarcane at optimum schedule of irrigations (based on IW/CPE ratio) at different locations

Location	*Soil type*	*No. of irrigation (depth, cm)*	*Optimum IW/CPE*	*Grain yield (t/ha)*	*Irrigation water requirement (cm)*	*Water use (cm)*
Cotton						
Sriganganagar (Raj.), *Kharif*	Sandy loam	8 (7)	0.75	2.25	56.0	99.0*
Parbhani (MS.), *Kharif*	Medium deep clay	6 (6)	0.90	1.64	36.0	74.1*
Bhatinda (Punjab), *Kharif*	Sandy loam	6 (7.5)	0.90	2.63	45.0	66.0*
Navsari (Gujarat), *Rabi*	Clay	9 (8)	0.75	1.28	72.0	74.7
Dharwad (Karnataka), *Kharif*	Med. black clay	5 (6)	0.60	1.41	30.0	68.0*
Hyderabad (A.P.), *Kharif*	Black clay	9 (5)	0.60	1.70	45.0	60.4*
Rahuri (MS.), *Kharif*	Black clay	9 (6)	0.75	2.64	54.0	102.0*
Bhavanisagar (T.N.), *Kharif*	Clay loam	6 (5)	0.60	2.02	30.0	64.0
Sugarcane						
Sriganganagar (Rajasthan)	Sandy loam	21–24 (7.5)	1.20/1.25	99.2	168.8	226.7*
Srigarganagar (Rajasthan)	Silt loam	19–24 (7.5)	1.20/1.25	110.0	169.0	213.0
Rahuri (Maharashtra)	Clay	32–35 (7.5)	1.0/1.05	128.5	252.0	336.8*
Parbhani (Maharashtra)	Med. black clay	26–32 (7)	1.0/1.05	82.6	206.5	253.9*
Parbhani (Maharashtra)	Deep clay	30 (7)	1.0/1.05	151.3	210.0	332.0
Navsari (Gujarat)	Deep black clay	7 (6)	0.75/0.80	68.4	42.0	166.4*
Pusa (Bihar)	Silt loam	4 (6)	0.60/0.65	76.7	24.0	97.9
Ludhiana (Punjab)	Sandy loam	9 (8)	1.0/1.05	87.1	72.0	207.0
Bhavanisagar (Tamil Nadu)	Sandy loam	17 (7)	0.75/0.80	94.1	119.0	193.5*

*Includes one or two irrigations of prescribed depth, ** Includes 4 irrigations of 7.0 cm depth @Seed cotton yield in case of cotton and millable cane yield in case of sugarcane

Source: Annual Reports of All India Coordinated Research Project on Water Management for 1972–73 to 1996–97.

Table 4.42: Grain yield, irrigation water requirement, and water-use in major pulse crops at optimum schedule of irrigations (based on IW/CPE ratio) at different locations

Location/ Season	*Soil type*	*Season*	*No. of irrigation*	*Optimum IW/CPE*	*Grain yield (t/ha)*	*Irrigation water*	*Water use*
Pigeon Pea							
Morena (M.P.)	Sandy loam	*Kharif*	3(7.5)	0.90	2	22.5	
Kota (Rajasthan)	Clay	*Kharif*	4(6)	0.75/0.80	1.45	24	
Navsari (Gujarat)	Clay	*Kharif*	2(8)	0.60	2.06	16	
Rahuri (MS.)	Clay	*Kharif*	6(7)	0.75/0.80	4.35	42	60
Green Gram							
Pantnagar (U.P.)	Silt clay loam	*Summer*	5(6)	0.60	1.42	30	38.2*
Dharwad (Karnataka)	Heavy clay	*Kharif*	1(6)	0.60	1.32	6	26.9*
Bhavanisagar (T.N.)	Silt loam	*Summer*	5–6(4)	0.60	1.32	22	26.5
Black gram		*Kharif*	7–8(4)	0.60	1.27	30	32.5
Bhavanisagar (T.N.)	Sandy loam	*Kharif*	4(9)	0.60	1.14	36	45.8
Chalakudy (Kerala)	Silt loam	*Summer*	5(3)	0.75	0.65	15	20.6*
Kidney bean							
Pusa (Bihar)	Sandy loam	*Rabi*	2(6)	0.80	1.19	12	21.8*
Faizabad (UP.)	Silt loam	*Rabi*	3(5)	0.70	1.56	15	19.8
Lentil							
Pusa (Bihar)	Sandy loam	*Rabi*	2(6)	0.60	1.60	12	15.8
Jabalpur (M.P.)	Heavy clay	*Rabi*	3(7.5)	0.60	1.62	22.5	29.3
Bikramganj (Bihar)	Sandy loam	*Rabi*	3(6)	0.60	0.98	18	24.9

Source: Annual Reports of All India Coordinated Research Project on Water Management for 1972–73 to 1996–97

in canal distribution systems reduce crop productivity and soil fertility in many canal commands. The water table rise ranges from 0.26–1.2 m in different study commands (Table 4.43). Once the water table rises to shallow depths, the land loses productivity due to unfavorable conditions like poor aeration in the root zone causing obstruction to germination and crop growth, reduction in activity of soil micro organisms, hindrance in farm operations, waterlogging and upsetting the salt-water balance leading to development of soil salinity.

Table 4.43: Water table rise in different canal command

Name of the irrigation command	*Rate, m/year*
Mahi Right Bank Canal Command, Gujarat	0.28
Rajasthan Canal Command, Rajasthan	029–0.88
Western Jamuna & Bhakra Canal Command, Haryana	0.3–1.00
Sirhond Canal Command, Punjab	0.10–1.00
Sharda Sahayak Canal Command, Uttar Pradesh	0.68
Malaprabha Canal Command, Karnataka	0.60–1.2
Nagarjuna Sagar Irrigation Project, Andhra Pradesh	0.32
Sriram Sagar Irrigation Project, Andhra Pradesh	0.26

Source: Ministry of Agriculture, GOI.

4.5.3.2 *Waterlogging*

Imperfect drainage and non-judicious use of water have resulted in the problem of water logging in many productive areas. In fact, salinity and waterlogging are the inter – linked problems. Water use efficiency of many command areas are affecting due to waterlogging and poor drainage. In the Tungabhadra Project of Karnataka more than 33,000 ha area are waterlogged. Similar situation exists in other command areas too. As per the report of National Commission on Agriculture (1976) the waterlogged areas in some major states in the country is 39.93 lakh hectors (Table 4.44).

4.5.3.3 *Soil salinisation*

The rise of water table in semi-arid and arid areas may mobilize the salts present in soil profile and ground water. Once the ground water table rises within 2–3 m of the soil surface, it contributes substantially to evaporation from the soil surface and water uptake by plants. The upward flux of water results in gradual increase in concentration of

Table 4.44: Waterlogged area of major states

State	*Area (lakh ha.)*
Andhra Pradesh	3.39
Bihar	1.17
Gujarat	4.84
Haryana	6.20
Punjab	10.9
Rajasthan	3.48
Uttar Pradesh	8.10
West Bengal	1.85
Total	39.83

Source: Ministry of Agriculture, GOI

salts in the shallow depths of soil profile. In the initial years, crop yield may be reduced due to salinity, but as the severity of the problem increases, the lands may have to be abandoned since the excess salts render the soils completely unproductive. Soil salinity problems develop whenever soil and hydrological conditions favour the accumulation, of soluble salts in the root zone. For formation of saline soils, a source of salt is necessary. The common sources are the salts present in the soil and ground water and those applied through irrigation. The source of salts in the soil could be that transported along with the alluvium and or that accumulated from weathering of the soils. These soils generally do not occur extensively in humid areas, where the salts get continuously leached out of the soil by excessive rainwater. Saline soils, however, are quite common in the semi-arid and arid regions receiving inadequate precipitation to accomplish leaching of salts originally present in the soil profile. Table 4.45 shows the estimates of saline areas in India.

Table 4.45: Estimates of saline areas in India

Source	*Saline areas (Mha)*
Chauhan. 1996. Technical Papers MOWR, New Delhi	7.2
Sewa Ram. 1996. Technical Papers MOWR, New Delhi	3.3
Abrol I.P. 1994. Salinity Management for Sustainable Agriculture, CSSRI, Karnal	10.9
Singh N.T. 1994. Natural Resources Management for sustainable use	8.6
Singh N.T. & Bandyopadhay, 1996. Soil Management in Relation to Land Degradation & Environment, IARI, Delhi	8.6
Sehgal J. 1996	10.1

Source: Hand Book of Agriculture 2009. ICAR New Delhi.

4.5.3.4 *Lowering of water table in well irrigated areas having good quality ground water*

While main problem in canal command is rise in water table, waterlogging and salinity, in well irrigated areas the problem is the receding groundwater due to over draft. In many states groundwater table is falling very fast at an annual rate of 1 meter. The excessive mining is observed in Punjab with a growth rate of 12% followed by Rajasthan, Tamilnadu and Haryana with a percentage of around 1–4. In Punjab, Haryana, Rajasthan fast receding water table contributes to water salinity problems and in peninsular India its major effect is that many wells are becoming dry. In Gujarat, it is a combination of both, with the added problem of fluoride contamination. In West Bengal, the arsenic toxicity is observed as a result of over draft. In the coastal belt, it results in the ingress of sea water.

As per the AICRP on Ground Water Utilization, 2009–10, the area of the Punjab state under water table depth greater than 10 m has increased from 0.95 M ha to 2.70 m ha indicating fall of groundwater beyond critical limit of 10 m, occurred mostly in areas that were earlier under the safe limit *i.e.* 3–10 m (Table 4.46).

Table 4.46: Area ('000 ha) under different water depths (m) for state of Punjab

Year	*0–3*	*3–10*	*10–20*	*>20*
1998	239.41	3776.50	962.06	40.34
1999	206.89	3837.52	937.10	36.81
2000	165.91	3561.05	1264.63	26.73
2001	162.76	3224.43	1595.57	35.55
2002	142.71	2966.11	1863.86	45.64
2003	24.58	2587.64	2397.14	8.95
2004	68.21	2362.10	2532.92	55.09
2005	27.36	2214.46	2679.17	97.33
2006	108.04	1665.98	2724.05	190.24

Source: Annual Report, AICRP on Ground Water Utilization, 2009–10

4.5.3.5 *Water quality deterioration in coastal areas*

While indiscriminate draft of ground water in inland areas leads to lowering of water table and increasing the cost of irrigation, the problem is quite different in coastal regions. Decline in water table owing to excessive pumping in coastal areas leads to ingress of sea water into the ground aquifers. Along the Saurashtra coast, 160 km long and 5–10 km wide ground water belt has been salinized due to overdraft of fresh water floating over saline aquifers. Thousands of wells in this area have gone out of commission or the lands have been salinized where water from such wells was continuously used for irrigation. Sub-terranian sea water intrusion has also been observed along the east coast in Orissa and Andhra Pradesh (Singh and Bandyopadhyay, 1996). Though in many places efforts are being made to replace saline water with fresh water ponded from local rivulets, but the process of replacement is slow because of differences in density of fresh and saline waters.

4.5.3.6 *Nitrate, fluoride and arsenic pollution of ground water*

Pollution of ground water, owing to leaching of nitrates, is also a concern in India. However, in general, at such low rates of N-fertilizer use prevalent in the country, nitrate leaching is not likely to pose serious problems in most of the farming situations. An increase in NO_3-N content of well waters has been registered only in the areas, where heavily fertilized and irrigated cropping systems predominate. In Ludhiana district of Punjab, average NO_3-N content of shallow wells increased from 0.42 to 2.29 mg^{-1} in the last century. Bajwa *et al.* (1993) analyzed water samples of 144 tube wells serving intensive cropping systems and noticed that percentage of samples having more than 5 mg NO_3-N was highest under vegetable cultivation, followed by potato-wheat and rice-wheat systems. Fertilizer management practice, season, frequency and depth of irrigation, percolation rate of the soil and rooting pattern of the crops, are most important factors influencing nitrate leaching (Singh, 1996). Another serious health hazard facing the country is the prevalence of disease known as fluorosis, which arises primarily due to excess of fluoride in natural water. The occurrence of arsenic in natural water of West Bengal and Bangladesh has received significant attention during recent years. In India, the contamination of ground water by arsenic was first detected in the state of West Bengal in 1978 and toxicity has multiplied due to over exploitation of ground water resources.

4.5.4 Strategies to Enhance Water Productivity, Water use and Irrigation Efficiency

Strategies to enhance water use efficiency and irrigation efficiency should include improved irrigation system management to provide more reliable water supply to farmers through storage and improved operation of reservoirs, better distribution of water with improved control structures as well as more responsive management. More reliable water supply allows farmers to invest in better on-farm water management such as better land leveling, zero tillage, or pressurized irrigation. Improved management usually requires improved institutions as well as improved technologies. Irrigation efficiency mainly depends on loss of water in conveyance system including distribution system and in the field. Loss of water in conveyance system is mainly due to seepage and evaporation. Evaporation loss takes place from the exposed water surface. Some of the strategies are discussed here to improve water and irrigation use efficiency.

4.5.4.1 *Multiple use of water for enhancing water productivity*

Per person availability of land, water and other natural resources is already low and is further declining. Multiple use of water through agricultural diversification which has several components like poultry, duckary, fisheries, apiculture etc. along with diversified crop production or value added products could be proved as effective means in mitigating drought and eliminating poverty as it provides high income, regular employment and balanced and quality food, even with less water. As the agricultural diversification comprises of several farm enterprises in addition to diversified crops, it provides opportunity to farmers to have a 'basket of complementary options' for reducing the weather risks involved in single commodity based agriculture.

Complementarities and synergies of raising crops, horticulture, livestock, aquaculture and micro-enterprising for enhanced returns, income and employment is becoming essential. Integrated production systems are based on cycling, recycling, multiple uses of inputs and reduced environmental loading. The enhanced income and employment is generally staggered in the enabling farming systems for ensuring continuous cash flows throughout the year.

Study revealed through cultivation of fishes, *rabi* crops with the harvested water of rainy season and on-dyke horticulture enhanced the water productivity from in rice Rs. 1.22/m^3 to Rs. 6.26/m^3 in integrated farming system (Fig. 4.18).

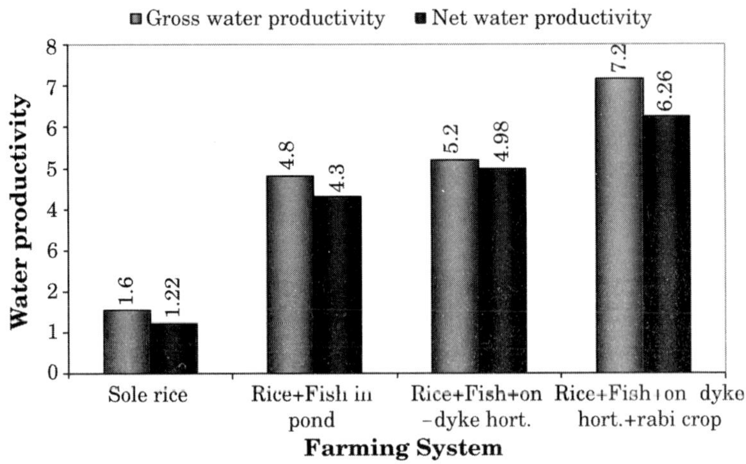

Fig. 4.18: Comparison of water productivity from different components of farming system. *Source*: Kar *et al.*, 2010

4.5.4.2 *Conjunctive use of surface and ground water*

Conjunctive use of Water

Conjunctive use of multiple water resources should be practiced to mitigate the effect of the shortage in canal water supplies often subject to steep variations in river flow during different periods in the year, to increase the dependability of existing water supplies, to alleviate the problems of high water table and salinity resulting from introduction of canal irrigation, to facilitate the use of high salinity ground waters which, otherwise, cannot be used without appropriate dilution.

The optimum conjunctive use of the regions through surface and groundwater resources would help in minimizing the problem of waterlogging and groundwater mining. The conjunctive use also facilitates the use of highly saline groundwater which cannot be otherwise use without appropriate dilution. Strengthening of knowledge based on geology and aquifer characteristics, hydrology of surface and groundwater facilities is required to develop appropriate conjunctive use system.

4.5.4.3 *Ground water recharge*

Ground water played a major role in the success of green revolution and contributes 60% of the total irrigated area of the country. Over exploitation of groundwater has reached danger level in Haryana, Punjab, and Rajasthan and Tamilnadu. Table 4.47 shows the groundwater development in different states. The Punjab-Haryana region could lose its production potential in a few decades if current pattern of groundwater extraction and pollution, soil salinization and rice wheat cropping system persists. Necessary steps must be taken for groundwater recharging in these areas. Rainwater harvesting has considerable potential for recharging depleted groundwater aquifers meeting irrigation demand.

4.5.4.4 *Irrigation scheduling based on scientific approach*

Irrigation scheduling is a systematic method of deciding the quantity and timing of irrigation. It helps the irrigator to decide on when and how much water to apply for minimizing crop yields and efficiency of water use. The basic objective of irrigation scheduling is to make available the correct amount of water for the biological processes of plants at appropriate time by applying the exact amount of water needed to replenish the soil moisture to the desired level. Irrigation scheduling becomes particularly sensitive under scare water supplies where water shortage requires a refined timing of water application in order to minimize yield restrictions.

4.5.4.5 *Pressurized irrigation system*

The water use efficiency under conventional flood method of irrigation, which is predominantly practiced in Indian agriculture, is very low due to substantial conveyance and distribution losses. Recognition the fast decline of irrigation water potential and increasing demand for water from different sectors, a number of demand management strategies and programme have been introduced to save water and increase the existing water use efficiency in Indian agriculture. One such method, introduced relatively recently in Indian agriculture, is pressurized irrigation system, which includes both drip and sprinkler method of irrigation. Since, these methods need water under pressure; they are classified as pressurized irrigation system. Pressurized

Table 4.47: Groundwater potential in the states & union territories of India (cubic km/year)

States/U.T.	*Total replenishable groundwater resource*	*Provision for domestic industrial and other uses*	*Available groundwater resources for irrigation*	*Net draft*	*Balance groundwater potential available for exploitation*	*Level of groundwater development (%)*
Andhra Pradesh	36.50	2.67	32.95	14.90	17.65	45
Arunachal Pradesh	2.56	0.009	2.30	0.0008	2.29	0.04
Assam	27.23	0.98	24.89	5.44	19.06	22
Bihar	29.19	2.14	27.42	10.77	15.89	39
Chhattisgarh	14.93	0.70	13.68	2.80	10.67	20
Delhi	0.30	0.57	0.28	0.48	0.00	170
Goa	0.28	0.04	0.27	0.07	0.18	27
Gujarat	15.81	1.48	15.02	11.49	3.05	76
Haryana	9.31	0.60	8.63	9.45	–1.07	109
Himachal Pradesh	0.43	0.04	0.39	0.12	0.25	30
Jammu & Kashmir	2.70	0.42	2.43	0.33	1.92	14
Jharkhand	5.58	0.56	5.25	1.09	3.99	21
Karnataka	15.93	1.41	15.30	10.71	6.48	70
Kerala	6.84	1.40	6.23	2.92	3.07	47
Madhya Pradesh	37.19	1.74	36.33	17.12	17.51	48
Maharashtra	32.96	1.52	31.21	15.09	16.10	48
Manipur	0.38	0.02	0.34	0.002	0.31	0.65
Meghalaya	1.15	0.10	1.04	0.002	0.94	0.18
Mizoram	0.04	0.0008	0.04	0.0004	0.04	0.90
Nagaland	0.36	0.03	0.32	0.009	0.30	3
Orissa	23.09	1.22	21.01	3.85	18.78	18

Table 4.47: (*Contd...*)

Table 4.47: (*Contd...*)

States/U.T.	*Total replenishable groundwater resource*	*Provision for domestic industrial and other uses*	*Available groundwater resources for irrigation*	*Net draft*	*Balance groundwater potential available for exploitation*	*Level of groundwater development (%)*
Punjab	23.78	1.00	21.44	31.16	–9.89	145
Rajasthan	11.56	2.72	10.38	12.99	–3.94	125
Sikkim	0.08	0.02	0.08	0.01	0.05	16
Tamil Nadu	23.07	0.91	20.78	17.65	3.08	85
Tripura	2.19	0.20	1.97	0.17	1.89	9
Uttar Pradesh	76.35	5.30	70.18	48.78	19.52	70
Uttarakhand	2.27	0.08	2.10	1.39	0.68	66
West Bengal	30.36	1.24	27.48	11.65	15.32	42
All States Union Territories	432.42	29.12	398.70	230.44	161.92	58
Andaman & Nicobar	0.330	0.008	0.320	0.010	0.303	4
Chhattisgarh	0.023	0.000	0.020	0.000	0.020	0
Dadra & Nagar Haveli	0.063	0.008	0.060	0.009	0.051	14
Daman & Diu	0.009	0.003	0.008	0.009	–0.002	107
Lakshyadweep	0.012	–	0.004	0.002	–	63
Pondicherry	0.160	0.031	0.144	0.151	–0.008	105
Total UTs	0.597	0.050	0.556	0.181	0.366	33
All India	433.02	29.17	399.25	230.62	162.29	58

Source: Hand Book of Agriculture, ICAR, 2009

irrigation system is proved to be an efficient method in saving water and increasing water use efficiency as compared to the conventional surface method of irrigation, where use efficiency is only about 35–40% (Table 4.48).

Table 4.48: Irrigation efficiencies in pressurized irrigation system

Factors	*Sprinkler irrigation system*	*Drip irrigation system*	*Surface irrigation system*
Overall irrigation efficiency	50–60%	80–90%	30–35%
Application efficiency	70–80%	90%	60–70%
Water saving	30%	60–70%	N.A.

4.5.4.6 *Irrigation layout*

It has been observed that farmers don't adopt the suitable layout as per the crop requirements. Generally restricted flooding is followed. But irrigation layout of appropriate sizes are very much essential for proper distribution of irrigation water, encourages uniform crop growth, reduces water losses, helps in soil and water harvesting along with size of irrigation streams, depth of irrigation, soil infiltration rate, land scape, method of irrigation and crop to be irrigated are the factors which decides the size of the individual farm plots to secure uniform water application without wastage or loss of nutrients. The length of the field should not exceed 90 m and width should be less than 15 m.

4.5.4.7 *Improving water use efficiency of canal command*

Considerable amount of water is lost in canal command through evaporation, seepage and percolation. In no irrigation project in India, the total losses in the canal distribution system and field has been less than 50% of the head discharge. A review of 90 irrigation projects of the World indicated generally low irrigation efficiencies, with only 20–40% of water diverted from the reservoir being effectively used by the crop, while in India, the irrigation efficiency is around 10–20%. The losses of irrigation water are unlined canal distributor system in north India is given in Table 4.49. The method to reduce conveyance losses are by lining the bed and sides of the canal, eradication of weeds, reduction of wastage from escapes and tail ends. The use of plastic films as lining material has offered tremendous scope in India.

Table 4.49: Losses of irrigation water in unlined canal distribution systems and in the field

Source of loss	*% of supplies at canal head*		
	Seepage	*Evaporation*	*Total*
Main canals and branches	13.6	3.4	17.0
Distributaries (10% of supply at distributory)	6.4	1.6	8.0
Field water courses (27% of supply at outlet head)	16.0	4.0	20.0
Losses from field during application (30% of supply reaching the head)	13.2	3.3	16.5
Total	49.2	12.3	61.5

In any irrigation project the total losses in the canal distribution system and field has not been less than 50% of head discharge. Studies conducted in Nagarjuna Sagar Project have revealed that only 40–60% of water released from the reservoir reaches the field and only 20–40% is used by the crops.

Irrigation efficiency is considered to be as low as 10–20% in some projects in India. It is estimated that 11 million ha of water is contributed to ground water recharge annually through seepage from canal system and the return flow from irrigation. To increase the overall irrigation efficiency of 40–50% for a project, improved irrigation and drainage practices should be adopted based on scientific land water management principles.

4.5.4.8 *Water saving technologies in canal command*

Irrigated agriculture can be made vibrant by increasing farm productivity in the country through scientific and efficient management of canal systems and water production can be enhanced through adoption of water saving technologies. Prominent water saving technological interventions taken by Directorate of Water Management, Bhubaneswar are alternate raised and sunken bed technique in low lands, system of rice intensification (SRI) technique of rice cultivation, paired row planting of groundnut, vermicompost application and plastic mulching in vegetable crops, etc.

Rotational irrigation is often recommended to irrigate a large area with a limited water supply and to ensure better equity among water users. Rotational irrigation is the application of required amounts of water to fields at regular intervals. The field may often be without standing water between irrigations, but ideally the soil does not dry enough for moisture stress to develop. A major advantage of rotational irrigation is possibly the more effective use of rainfall.

Plant height, tiller number, leaf area index, dry matter production and grain yield generally decrease as the irrigation interval increases from 4 to 10 days (Table 4.50).

Table 4.50: Effects of rotational irrigation in IR 20 rice

Irrigation interval (days)	*Plant height (cm)*	*Tillers (No. hill)*	*LAI*	*Total dry matter (g/hill)*	*Yield (t/ha)*
4	102	18	6.5	53	7.2
6	97	18	54	47	7.1
8	98	17	5.2	49	6.8
10	92	12	4.2	37	5.6

Source: Reddy and Reddy, (2009)

The results of experiments conducted by WALAMTARI on rice in Andhra Pradesh under Sriramasagar Project in sandy loam soils are given in Table 4.51.

Table 4.51: Saving in water due to rotational irrigation for rice crop

Irrigation interval (days)	*Quantity of water applied (mm)*	*Yield of rice (kg/ha)*	*Reduction of yield (%)*	*Water saved (%)*
1	1717	6324	–	–
2	1117	6056	4.42	34
3	1027	5924	6.75	40
4	913	5720	10.55	46
5	877	5646	12.00	48

Source: Reddy and Reddy, (2009)

4.5.4.9 *Re-use of irrigation water*

About 30% of water applied to the field is lost as surface or sub-surface flow and reaches drains. The water that reaches the drains can be stored at suitable sites and pumped back for irrigation.

4.5.4.10 *Optimum design of farm irrigation systems*

Farm irrigation systems are designed to match the physical and economic conditions in which they are to operate. The physical factors that influence the design of irrigation systems are: climate, soil, topography, location of irrigation source etc. The economic factors are cost of land, the existence of market, production cost, cost of the energy and water, crop prices, etc. The important steps in designing farm irrigation systems are collection of data (weather, crop, soil, hydrology) needed for design, evaluation of water Source (tanks, canals and wells, water quality), design daily irrigation requirement (ratio of allowed depletion of soil water before irrigation (mm) to minimum irrigation interval (days), preparation of alternate designs and assessing performance of farm irrigation systems. Cost of the project and availability of funds are two important factors in selecting the irrigation system. Based on all the factors, the most suitable system is selected.

4.5.4.11 *Better operation planning*

Irrigation system operation is the process of releasing, conveying and diverting water in the canal systems to ensure pre-determined flows at prescribed times for specified duration at all designated points of delivery. The operation plan takes into account the water available in the reservoir at the beginning of each crop season and spells out the starting date of release of water, the mode of supplies *i.e.* whether intermittent and continuous, the detailed schedule of releases and the closing date of release of water.

The operation plan is prepared with active participation and involvement of farmer's representatives so that it may be implemented without any resistance from the farmers to plan their crop production operations properly. This will reduce wastage and better utilization of irrigation water will be ensured.

4.5.4.12 *Systematic canal operation*

The object of systematic canal operation is to ensure equitable distribution of available flows at the heads of all off-takes on a distributory. A schedule of canal operation has to be prepared in advance. For instance, a distributory may be divided into seven reaches each covering one or more off-takes whose total discharge is about a seventh of the designed discharge of the distributory. One-seventh of the total discharge can be saved on each day by closing the off-take in each reach for 24 hours in succession in a week. This water can be made available to tail-enders and thus it will help saving water besides not having dry tail-enders problem.

4.5.4.13 *Response to rainfall in canal operations*

Canal operations should respond meticulously to rainfall in the command area for efficient use of storage water in the reservoir. The demand of crops for water is met in the wet season by both rainfall in the command area and water stored in the reservoir. Water requirements of crops are worked out using one of the empirical methods suggested by FAO, Rome. The most common method used is Penman-Monteith (1996) method. While assessing the water requirements of crops during the wet season for preparing the operation plan, effective rainfall worked out from the data of average rainfall over a period of past 20 years is deducted from the water requirements of crops worked out by the Penman-Monteith (1996) method or other empirical methods. But in real time situations, the pattern and amount of rainfall varies, influence the scheduled deliveries. When rainfall is in excess, the scheduled irrigations are skipped and when rainfall is deficient unscheduled deliveries have to be made. This will ensure maximum utilization of rainfall as well as protecting the crops from water stress or from excess water.

Irrigation deliveries are based mainly on the status of water balance in the cropped fields. The water balance in paddy fields is on the basis of ponding depth while for light irrigated crops it is on the basis of water storage in the root zone. The allowable depletion of soil water storage is limited by readily available soil moisture capacity. Response to rainfall for paddy depends on adopted ponding depth while for light irrigated crops it depends on the depletion level of readily available soil water from the root zone. The mode of irrigation supplies differ during various stages of crop growth and development, as water requirements of crops vary at different stages of crop growth and development.

4.5.4.14 *Precision land leveling*

Declining irrigation water availability, sustainable crop production in increasing food demand necessitates quick adoption of modern scientific technologies for efficient water management. Land leveling at farmer's field is an important process in the preparation of land. It enables efficient water utilization of scarce water resources through elimination of unnecessary depression and elevated contours. Laser leveling is laser guided precision leveling technique used for achieving very fine leveling with desired grade on agriculture field. It is becoming popular in almost entirely state of Punjab. Precision land leveling must be treated as a precursor technology for improving crop yields, enhancing input-use efficiency and ensuring long term sustainability of the resources base in intensively cultivated areas. Keeping in view the benefits of laser leveling, the study was conducted to evaluate the performance laser leveler at farmer's field and to estimate the saving in water resources at different level of adoption. Water saving in different crops at different level of laser leveler is presented in Table 4.52.

Table 4.52: Water saving in different crops at different level of laser leveller

Crop	*Water saved (%)*	*Water required (cm)*	*Water saved (cm)*	*Area (ha)*
Maize	27.1	35	9.5	152560
Wheat	26.0	35	9.1	3469520
Cotton	27.25	20	5.5	541060
Paddy	26.33	160	42.1	2625204
Total	26.25	81.3	21.3	6788344

Crop	*Water saved (ha-m) at different level of irrigation*				
	10	*25*	*50*	*75*	*100*
Maize	1447	3618	7235	10853	14470
Wheat	31573	78932	157863	236795	315726
Cotton	2949	7372	14744	22116	29488
Paddy	110595	276486	552973	829459	1105946
Total	146563	366408	732816	1099224	1465631

Source: Annual Report of All India Coordinated Research Project on Water Management, 2009–10

4.5.4.15 *Management of poor quality groundwater*

Groundwater in arid regions is largely saline and in semi arid it is sodic in nature. Indiscriminate use of poor quality water for irrigation deteriorates productivity of soils through salinity, sodicity and toxic effects. Table 4.53 shows the use of poor quality groundwater in different states. In addition to reduced productivity, it deteriorates the quality of produce and also limits the choice of cultivable crops. In saline groundwater areas, where canal water is available in limited quantity, conjunctive use of canal water and saline groundwater is recommended. Dilution or mining of available poor quality water with good water should be done in such proportion that resultant EC is acceptable for the range of crops to be grown in a given area. Some of the management options to manage the saline groundwater are selection of semi tolerant crops, salinity tolerant cultivars, proper selection of crop sequence, avoiding saline water use during initial growth stages, addition of farm yard manure, organic matter, application of additional dose of nitrogen etc. Some important points should be noted for management of conjunctive use practices such as analysis of saline water to evaluate its use potential, selection of crops/ varieties, pre sowing irrigation with good quality water so that germination and seedling emergence are not affected, adequate leaching of salts, change of cropped area and adoption of improved cultural and nutrient management practices.

Management technologies options for use of sodic groundwater includes selection of crops, appropriate irrigation scheduling and conjunctive use options with canal water, river water management, leaching strategies to maintain a high level of soil moisture and low level of salts and exchangeable sodium in the rhizosphere, use of land management practices to increase the uniformity of water distribution, infiltration and salt leaching besides the optimal use of chemical amendments like agricultural grade gypsum and acidic pyrile at proper time and mode of their application with judicious use of organic materials and chemical fertilizers.

4.5.4.16 *Mitigating groundwater pollution*

The choice of appropriate cropping systems and management practices helped minimizing nitrate leaching besides improving N-use efficiency.

Table 4.53: Use of poor quality groundwater in different states

State	*Utilizable groundwater*	*Net draft*	*Groundwater development (%)*	*Poor quality water use (M ha-m/yr)*	*Saline groundwater area >4dSm-1 (km²) (M ha-m/yr)*
	(M ha-m/yr)				
Punjab	1.47	1.67	98	0.68	3058
Haryana	0.86	0.72	76	0.47	11438
Uttar Pradesh	6.31	2.98	42	1.42	1362
Rajasthan	0.95	0.77	73	0.65	141036
Bihar	2.06	0.82	36	NA	NA
W. Bengal	1.77	0.63	32	NA	NA
Delhi	0.01	0.01	120	NA	NA
Gujarat	1.56	0.85	49	0.26	24300
Karnataka	1.24	0.45	33	0.17	8804
Tamilnadu	2.02	1.40	63	NA	3300
Madhya Pradesh	2.66	0.73	25	0.20	NA
Maharashtra	2.29	0.88	35	NA	NA
Andhra Pradesh	2.70	0.78	26	0.25	NA
India	32.63	13.50	37		193438

Source: Minhas *et al.* (2004)

Legume intercropping in cereals grown with wider row spacing has been reported to reduce nitrate leaching (Yadav, 1981). Parallel multiple cropping (a system of growing two dissimilar growth habit crops with minimum competition) of sugarcane and blackgram and that of pigeonpea and maize resulted in low NO_3-N content in soil profile as compared to sole cropping (Yadav, 1982). As a crop management strategy to minimize NO_3-N leaching, Singh *et al.* (1995) suggested delaying large N applications until the crop can utilize it and avoiding irrigation when large amount of NO_3-N present in the root zone.

There is a specific need to create awareness among population about the presence of fluoride in ground water, its action on body tissues and available remedial technologies. Various artificial recharge techniques including Aquifer Storage Recovery (ASR) technique may be applied to improve the quality of water by dilution. The ASR technique is being followed in many parts of the world and has proved to be a viable or cost-effective option for storing large volumes of fresh water. Looking at the success achieved through ASR in many countries, there is a case to plan trial of this technology at suitable locations in the country. Besides this, the concept of mixing two different aquifers/sources of water can also be tried to overcome the fluoride problem.

Investigations of geochemistry is very essential to understand the occurrence and mobilization of arsenic in the aquifer system. Thus, there is an urgency for an integrated study on geological, hydrological and geochemical characteristics of the multi-level aquifer system to predict the origin, occurrence and mobility of arsenic in ground water. There is a priority to arrive at suitable and economically viable alternatives to maintain sustainable drinking water supply. The traditional system of using surface run off water can be augmented by system to reduce dependence on ground water extraction.

4.5.4.17 *Farmers' participation in irrigation water management*

Farmers' active participation in irrigation water management is most essential as the farmers are the ultimate users of irrigation water and their participation will ensure successful operation and maintenance of the irrigation system.

4.5.4.18 *Integrated irrigation water management*

The process of water management in an irrigation system includes an integrated development of sub-systems such as watershed, reservoir intake, conveyance, regulation, measurement, effective use of rainfall, application of appropriate quantities for crop use and effective drainage from command area. For efficient use of irrigation water simultaneous development of all the components of an irrigation system is essential.

Although India has developed irrigation infrastructure for over 60 m ha, the use efficiency of irrigation systems is quite low leading to poor returns on investment and environmental problems like secondary salinity and water logging. Improving irrigation use efficiency and enhancing water productivity have now become very high priorities. Several technologies were developed for efficient water management in irrigated areas, but their farm level adoption has been quite low due to a number of socio-economic and policy constraints. Parihar (2006), outlined following key approaches for optimum use of limited irrigation water.

1. Reduce amount of irrigation water applied to all acres
2. Substitute low-water requirement crops for high-water requirement crops
3. Irrigate only during critical stages
4. Manage soil moisture to capture precipitation.
5. Reduce irrigated acreage for full supply of irrigation

Splitting fields between low and high water requirement crops will reduce total water needed and better distribute water use across the growing season. This strategy also benefits producers with low capacity wells. Delaying irrigation until critical times is also possible if water volume is limited, but well capacity is normal. Irrigation water must be allocated so that periods of possible water deficits coincide with the least sensitive growth periods. Thus irrigation scheduling should be decided based on the water availability. The procedure for optimal allocation of water under limited water supply condition includes quantifying water use (ET or T) vs. crop biomass relations and employing optimization models with operational constraints. Water availability during reproductive and grain fill growth stages is critical to grain production. During vegetative growth, some water stress can be tolerated without affecting grain yield and root development can be encouraged so the crop uses deeper soil water. Managing soil moisture to capture precipitation is important for all limited irrigation situations. Crop residues on the soil surface intercept rainfall and snow, enhance infiltration, and reduce soil evaporation.

4.5.4.19 *Better management techniques for higher water productivity*

Techniques for efficient management include conservation of water, integrated water use, optimum allocation, water efficient crops and enhancing water use efficiency. *In situ* conservation of water can be achieved by reducing runoff and enhancement of infiltration water and reduction of seepage losses and direct evaporation from soil. Water loss by deep seepage can be reduced by increasing soil water storage through enhancing the root zone of crop and soil water retention. Direct evaporation from soil can be controlled with shallow tillage and mulching. *Ex situ* conservation of water can be achieved by harvesting of excess water in storage ponds for its reuse for irrigation purpose.

Optimal allocation of available water among the competing crops and optimum timing of application is to be decided so as to maximize economic returns from available water. Under limited water supply situation, irrigation water must be allocated so that periods of possible water deficits coincide with least sensitive growth periods. Thus, irrigation scheduling should be decided based on the water availability. Water use efficiency by crops can be improved by selection of crops and cropping systems based on available water supplies and increasing seasonal evapo-transpiration (ET). The later can be achieved by selection of irrigation method, irrigation scheduling, tillage, mulching and fertilization. Selection of crops and cropping systems under limited irrigation situation should be such that there should be minimum water stress during the critical growth stages of crops although some water stress to the crops and associated yield reduction is inevitable.

Micro-irrigation methods *viz.*, sprinkler or drip systems for efficient utilization of available water in scarce areas, particularly for horticultural and plantation crops resulted in 30 to 50% saving of water and increased crop yields. Irrigation should be applied at critical crop growth stages to achieve maximum water use efficiency. The irrigation should be so managed that the inevitable stress synchronizes with the least sensitive stages. There are different methods of irrigation scheduling *viz.*, critical growth stage approach, soil moistures depletion approach, climatological approach (IW/CPE), leaf canopy temperature method which can be adopted for optimizing the timing of irrigation so as to reduce the water stress at critical growth stages.

Tillage, mulching, timely fertilization and moisture conservation by suppressing evaporative losses are other important techniques which save on irrigation water and improve its use efficiency. Regulating the depth of ponded water in crops like paddy, intermittent in place of continuous submergence are other ways of minimizing irrigation water. The over all benefits of improved water management technologies on wheat and rice yield and saving in irrigation water are given in Table 4.54. There is also need for identifying cropping systems that are more water productive in water scarce areas.

Table 4.54: Effect of water management technologies on yield and water saving in rice and wheat

	Yield, t/ha		*Water saving*	
	Farmers	*Improved*	*Farmers*	*Improved*
Rice	6.14	8.13	32.1	38
Wheat	3.20	4.17	30.3	42

4.5.4.20 *Policy initiatives*

With the availability of improved drilling technology, ground water irrigation became major source in Indian Agriculture. Subsidies on power, unrestricted use of ground water has resulted to excessive use of ground water resulting into further deepening of wells at many states. The government has initiated several measures like enhancing recharge, regulation of well/borewell digging. State Governments of AP and Maharashtra have enacted specific Acts like Water, Land and Tree Act. Gujarat government has initiated a project called "Jyotigram" to separate the power supply lines for domestic and agricultural systems at acost of US $290 million to cover 18,000 villages in about 4 years. Micro irrigation is being promoted at national level as major drive towards water saving. The Ministry of Agriculture and Ministry of Water Resources are implementing extensive capacity building programmes for extension officers and farmers on enhancing water productivity with a slogan of more crop per drop during XI plan.

Water forms the backbone for all the future endeavors to achieve the vision of food security. In the present data context, upscalling agricultural economic growth to more than 4% annually is the main challenge. Taking water technologies for better water management from lab to land is a formidable task to be addressed. Modernization / automation of irrigation system, precision irrigation, land reforms, corporate farming, cooperative farming, water and energy pricing, crop insurance, institutional mechanism for better governance, water rights are some of the key issues for better water management in agriculture. The projected food requirement demands a pronounced role for research, development and training in the water and agriculture sector.

It is evident that the water availability for agriculture is declining and to enhance agricultural production more water is needed. Therefore, concerted and holistic efforts are required in increasing the overall water use efficiency at system level which would be achieved through various measures like timely execution of projects, minimizing the losses, better operational efficiency through stake holders participation, implementation of on – farm water management technologies, conjunctive use of water and changes in irrigation policy. Simultaneously, the effort of R & D institutions are required in development of water management technologies, suitable database development, economic studies of various irrigation systems, policy guidelines for on farm water management and adoption of participatory irrigation management. The serious efforts of developmental agencies as well as research institute are required to develop a suitable water perspective plan for various regions in the country for its implementation.

4.6 DROUGHT MANAGEMENT AND CONTINGENCY PLANNING

Drought is one of the most important constraint affecting productivity and sustainability of agriculture. Since Indian agriculture is primarily monsoon dependent, deficit rainfall affects *kharif* crops; reduces the carry over soil moisture for *rabi* crops and affects livestock production through reduced fodder availability. Efficient management of natural resources is the most important strategy to mitigate droughts, followed by crop planning and economic/policy measures for relief.

4.6.1 Types of Droughts

Drought is a climatic anomaly, characterized by deficient supply of moisture resulting either from sub-normal rainfall, erratic rainfall distribution, higher water need or a combination of all three factors. Major parts of India have a probability of facing three to four drought years in a ten-year which could be of moderate intensity in one or two years severe in half to one year. The rainfed regions encompassing arid, semiarid and dry sub humid regions (covering regions less than 1150 to 1200 mm) are more prone to climate variability and drought (Ramakrishna *et al.*, 2007). Droughts are categorized in to meteorological, hydrological and agricultural. The criteria adopted by Central Water Commission (CWC) for drought identification is that when annual rainfall is less than 75% of the normal in 20% of the years. Agricultural droughts are however more relevant in the context of crop and livestock productivity. Depending on the time of occurrence, agriculture drought could be early season (due to delay in commencement of sowing rains), mid season (when there are breaks in southwest monsoon) or late season drought (due to early cessation of rains sometimes crops experience moisture stress its reproductive phase).

Chances of experiencing moderate droughts greater than 20 percent spread across greater part of the country whereas higher probabilities of experiencing severe drought are most confined to extreme northwest portion of Gujarat and Rajasthan (Table 4.55).

Table 4.55: Probability of occurrence of drought in different meteorological subdivisions of India

Meteorological subdivision	*Frequency of deficit rainfall (75% of normal or less)*
Assam	Once in 15 years
West Bengal, Madhya Pradesh, Konkan, Bihar and Orissa	Once in 5 years
South interior Karnataka, Eastern Uttar Pradesh and Vidarbha	Once in 4 years
Gujarat, East Rajasthan, Western Uttar Pradesh	Once in 3 years
Tamil Nadu, Jammu & Kashmir and Telangana	Once in 2.5 years
West Rajasthan	Once in 2 years

4.6.2. Impact of Droughts

The usual impact of droughts is in terms of loss of crops, malnutrition of human being and cattle, land degradation, loss of other economic activities, spread of diseases, and migration of people and livestock (Kulshrestha, 1997). Droughts result in crop losses of different magnitude, depending on their geographic incidence, intensity and duration. The droughts not only adversely affect the food supply at the farm level but also impacts national economy. Predicted losses to agriculture in India were 50 percent during the drought of 1957–58. Ramakrishna and Rao (1991) observed that during the 1987 drought in India, the productivity of pearl millet in western Rajasthan dropped by 78, 74 and 43 per cent in rainfall zones of < 300, 300–400 and > 400 mm, respectively. Victor *et al.* (1991) reported reduction in the productivity of groundnut and millet in Andhra Pradesh during 1987 drought year. For eastern India, the loss in production of food grains due to drought averaged over 1970–96 has been estimated to be $ 400 million, which is equivalent to 8 percent of the value of food grain production in the region (Pandey *et al.*, 2000). The effect of drought was more pronounced on fodder availability as compared to that of food grains. The food grain production was reduced by 29.0 and 17.6 mt during 2002 and 2009 drought years due to reduction in net sown area in *kharif* season compared to previous years that received good rains. Though the rainfall departure during 2009 was higher than in 2002, the fall in food grain production over the previous year was comparatively less in 2009 compared to drop in production (29 mt) in 2002. This has been mainly due to better contingency strategies planned and adopted during 2009–10 season and the greater resilience of Indian agriculture to drought now than in the past.

The duration of availability of water in surface water bodies was reduced significantly during 1987 drought year (Narain *et al.*, 2000). The water storage in major reservoirs were 33 percent less than the average of previous 10 years (Samra, 2004). Around 150 million cattle were affected due to lack of fodder availability. The groundwater table in drought affected areas was declined by 2 to 4 mts below the normal.

Droughts affect the livestock in several ways. Reduced productivity and mortality are the direct effects. Driven by enhanced livestock pressure due to depletion of forage resources during drought, overgrazing and indiscriminate cutting of vegetation takes place leading to land degradation. This is followed by first distress sale of cattle and even small ruminants. Migration is the next step extending the problems of uncontrolled and overgrazing, thus degrading land in other areas. There have been instances that large-scale mortality of livestock and mismanagement in disposal of their carcass caused epidemic situations and environmental hazards. Decrease in size of herd (up to 52%) was reported due to frequent occurrences of droughts in Rajasthan (Anonymous, 1994).

The impact of drought on agriculture and livestock production depends on the intensity and the time of occurrence in the season (Ramakrishna, 1997). In the arid regions, where mean annual rainfall is less than 500 mm, drought is inevitable phenomenon during most of the years. In semiarid regions, where mean annual rainfall varies between 500–750 mm, drought occurs in 40 to 60% of the years due to deficit seasonal rainfall or inadequate soil moisture availability between two successive rainfall events. Even dry sub-humid regions, where rainfall varies between 750–1200 mm, do experience intermittent drought situations due to break monsoon conditions. As a result drought management strategies have to be identified separately for each climatic region and also for major crops grown in those regions.

4.6.3 Management of Drought – Research & Development

4.6.3.1 *Research initiatives*

The Climatology Group at the Department of Meteorology, Andhra University, Visakhapatnam initiated studies on drought in 1960 using water balance approach and India Meteorological Department established a Drought Research Unit and classified droughts based on rainfall departures. Simultaneously, ICAR has also initiated work on impacts of droughts on agriculture through its Research Institutes, *viz.*, Central Soil & Water Conservation Training & Research Institute, Dehradun; Central Arid Zone Research Institute, Jodhpur; Indian Agricultural Research Institute, New Delhi and Central Research Institute for Dryland Agriculture, Hyderabad since their inception.

Keeping in view of the productivity gap between irrigated and rainfed areas, the Fourth Five Year Plan (1969–74) emphasized the need for extending the advancements in science and technology to dryland areas. More emphasis was laid on research related to dryland agriculture. The All India Co-ordinated Research Project on Dryland Agriculture (AICRPDA) was established at 24 centres in 1970 to tackle the problems of dryland agriculture, which are highly location-specific.

The ICAR has started an All India Coordinated Research Project on Agrometeorology, which is operating at 25 centres from 1995 onwards. These centres are providing weather-based agro-advisories to farmers in different agroclimatic regions of the country in addition to agroclimatic planning, contingency measures for different weather situations and disseminated through ICT facilities including mass media involving State Agricultural Universities as the main vehicle.

International Institutes like ICRISAT started in 1972 also helped semi-arid farmers in managing dryland crops, *viz.*, sorghum, pearlmillet, pigeonpea and groundnut for stable production under aberrant weather conditions. Enhancing production through watershed and farming system approach are strategies suggested by ICRISAT for climate risk management. A number of Research Institutes of ICAR and State Agricultural Universities are conducting considerable research work on developing drought and heat tolerant varieties and soil and water conservation measures, which help in combating drought and stabilizing crop production. A brief summary of these technologies is given in the following section.

4.6.3.2 *Strategies and technologies to combat drought*

Following are the various improved techniques and practices recommended for drought management in rainfed areas:

- ***Crop planning***: Crop varieties for dryland areas should be of short duration drought resistant/tolerant which can be harvested within rainfall periods and have sufficient residual moisture in soil profile for post-monsoon cropping.
- ***Crop substitution***: Traditional crops/varieties, which are inefficient utilizers of soil moisture (due to long duration), less responsive to external inputs and low yielders should be substituted by more efficient ones.

- ***Cropping systems***: Increasing the cropping intensity by using the practice of intercropping and multiple cropping results in more efficient utilization of resources. The cropping intensity would depend on the length of growing season, which in turn depends on rainfall pattern and the moisture storage capacity of the soils.
- ***Fertilizer use***: Dryland soils are not only thirsty, but also hungry. Therefore, application of fertilizers should be done in furrows below the seed. The use of fertilizers not only helps in providing nutrients to crop but also leads to efficient use of soil moisture. A proper mixture of organic and inorganic fertilizers improves moisture holding capacity of soils and enhances drought tolerance.
- ***Rain water management***: Efficient rainwater management can increase agricultural production from dryland areas. Application of compost and farmyard manure and raising legumes add organic matter to the soil and increase the water holding capacity. The water, which is not retained by the soil, flows out as surface runoff. This excess runoff water can be harvested *in situ* by proper land treatments or stored in dugout ponds and recycled to for supplemental irrigation.
- ***Watershed management***: Watershed management is a approach to optimize the use of land, water and vegetation in an area and thus, to provide solution to droughts, moderate floods, prevent soil erosion, improve water availability and increase fuel, fodder and agricultural production on a sustained basis.
- ***Alternate land use***: All drylands are not suitable for crop production. Some lands may be suitable for range/pasture management, ley farming, dryland horticulture, agro-forestry systems including alley cropping. All these systems, which are alternative to crop production, are called as alternate land use systems. These systems help to generate off-season employment in mono-cropped drylands and also, minimize risk, utilize off-season rains, prevent degradation of soils and restores balance in the ecosystem. Different alternate land use systems are alley cropping, agri-horticulture and silvi-pasture.

4.6.3.3 *Managing drought during different crop stages*

Specific recommendations to mitigate dry spells occurring in different crop stages are available for different regions of the country.

a. Early season drought

Early season droughts occur due to delay in commencement of sowing rains. Sometimes, early rains may occur tempting the farmers to sow the crops followed by a long dry spell leading to withering of seedlings and poor crop establishment.

The management options to cope up with early season drought are

- Raising community nurseries for cereal crops and transplant the seedlings with the onset of rainy season.
- Sowing of alternate crops/varieties depending upon the time of occurrence of sowing rains. The seeds may not be available to the farmers and government should provide seed through seed banks promoting seed villages.
- If there is a poor germination and inadequate plant stand, resowing is recommended. Seed priming helps in better crop establishment.

b. Mid-season drought

Mid-season drought occurs due to long gaps between two successive rainy events when the moisture available in the soil falls short of water requirement of the crop during the dry period.

The management options to cope-up with the mid season droughts are

- Rain water harvesting and recycling for life saving irrigation
- Reducing crop density by thinning
- Weed control and mulching
- *In-situ* moisture conservation
- Dust mulching, repeated harrowing
- Use of anti-transpirants

c. Late season drought

If the crop encounters moisture stress during the reproductive stage due to early cessation of rainy season, there may be rise in temperature, hastening the process of crop maturity. The grain yield of crops is highly correlated with the water availability during the reproductive stage of growth. Short duration high yielding varieties may escape late season droughts. Another possibility is to provide supplemental irrigation from harvested rainwater. Organic mulches are found to be useful in improving crop yields during post-rainy season. When crops are sown late, terminal drought can be anticipated with greater certainty. Therefore, varieties that respond better to terminal droughts have to be preferred.

4.6.3.4 *Soil and water conservation measures*

The Soil and Water Conservation (SWC) are the most important components of drought management (both temporary and permanent). SWC measures are essential for *in-situ* conservation of soil and water. The main aim of these practices is to reduce or prevent either water erosion or wind erosion, and enhancing moisture in soil for sustainable production. The suitability of *in situ* soil and water management practices depend greatly upon soil type, topography, climate, cropping system and farmers' resources.

Some of the most common *in situ* conservation practices followed in India are described below:

a. Contour cultivation

Contour cultivation is a simple method of cultivation across the slope which effectively reduces the runoff and soil loss on gentle sloping lands. In this, all field operations such as ploughing, planting and inter-cultivation are performed on the contour (Fig. 4.19). It helps in reduction of runoff by impounding water in small depressions and reduces the developments of rills. Maximum effectiveness of this practice is on medium slopes and on permeable soil.

b. Conservation furrow

Conservation furrow is a simple and low cost *in situ* soil and water conservation practice for areas with moderate slope. This practice is suitable for soils with problems of crusting, sealing and hard setting. In this system, series of furrows are opened on contour or across the slope at 1–3 m apart (Fig. 4.19). The spacing between the furrows and its size can be chosen based on the rainfall, soils, crops and topography. The furrows can be made either during planting time or interculture operation using country plough. Two to three passes in the same furrow may be needed to obtain the required furrow size. These furrows harvest local runoff and improve the soil moisture in the adjoining crop rows particularly during the period of water stress. This practice increases crop yields by 10–25 percent and it cost around Rs. 250-350 ha^{-1}. To improve its effectiveness further, it is recommended to use this system along with contour cultivation or cultivation across the slope (Ram Mohan Rao *et al.*, 1981; Mishra *et al.*, 2010).

c. Broadbed and furrow system

On black soils, the problem of water logging and water scarcity occurring during the same season are quite common. The *in situ* conservation practice should store water in the profile and drain excess water. The "Broadbed and Furrow" (BBF) system successfully meets these goals (Fig. 4.19). The BBF system consists of a relatively raised flat bed or ridge approximately 90 cm wide and shallow furrow about 30 cm wide and 15 cm deep. The BBF system is laid out on a grade of 0.4–0.8 % for optimum performance. It is important to attain a uniform shape without sudden and sharp edges because of the need to plant rows on the shoulder of the broadbed.

Contour cultivation *Conservation furrow* *Broad bed & furrow*

Fig. 4.19: Most common *in situ* conservation practices (a) contour cultivation (b) conservation furrow (c) broad bed and furrow system

d. Vegetative barriers

Vegetative barriers or vegetative hedges or live bunds are effective in reducing soil erosion and conserving moisture. In many situations, the vegetative barriers are more effective and economical than mechanical measures, *viz.*, bunding. Vegetative barriers can be established either on contour or on moderate slope of 0.4 to 0.8 percent.

In this system, the vegetative hedges act as a barrier to runoff flow and slowing down the velocity resulting in deposition of eroded sediments and increased rainwater infiltration. It is advisable to establish vegetative hedges on small bund. This increases its effectiveness, particularly during the first few years when the hedges are not so well established. If the main purpose of the vegetative barrier is to act as a filter to trap the eroded sediments and reduce the velocity of runoff, then grass species such as vetiver, sewan (*Lasiurus sindicus*), sania (*Crotolaria burhia*) and kair (*Capparis aphylla*) are recommended. If the purpose is to stabilize the bund, plants like Gliricidia could be used. Gliricidia plants grown on bunds not only strengthen the bunds, but also provide N-rich green leaf biomass, which can be added on soil as mulch-cum-manure.

4.6.3.5 *Water harvesting and ground water recharge*

While *in situ* conservation practices are appropriate at farm level, at community or landscape level, harvesting surplus runoff for surface storage or ground water recharge are more important measures for drought proofing.

a. Check dams

Masonry check dams are permanent structures used for controlling gully erosion, water harvesting and groundwater recharging (Fig. 4.20). These structures are popular in watershed programs in India. These structures are preferred at sites where velocity of runoff in gullies/ streams is very high and stable structure is needed to withstand the flow. Masonry check dams are designed on the basis of rainfall-runoff relationship. Depending upon the assumed depth of structure and the corresponding area to be submerged, suitable height of the dam may be selected to provide adequate storage in a given topographic situation (Katyal *et al.*, 1995).

Earthen check dams are very popular in the watershed programs in India for controlling gully erosion and for harvesting runoff water. These are constructed using locally available materials. The cost of construction is generally low. The size of the dam depends on the site conditions. In some cases, stone pitching may be required to protect the bund from scouring. Earthen check dams are used as surface water storage structures as well as for recharging groundwater.

b. Khadin system

'Khadin' is a land-use system developed centuries ago in the Jaisalmer district of Western Rajasthan. This system is practiced by single large farmer or by group of small farmers. It is highly suitable for areas with very low and erratic rainfall conditions. In khadin system, preferably an earthen or masonry embankment is made across the major slope to harvest the runoff water and prevent soil erosion for improving crop production. Khadin is practiced where rocky catchments and valley plains occur in proximity. The runoff from the catchment is stored in the lower valley floor enclosed by an earthen/stone 'bund'. Any surplus water passes out through a spillway. The water arrested stands in the khadin throughout the monsoon period. It may be fully absorbed by the soil during October to November, leaving the surface moist. If standing water persists longer, it is discharged through the sluice before sowing. Wheat, chickpea or other crops are then planted. These crops mature without irrigation. The soils in the khadins are extremely fertile because of the frequent deposition of fine sediment, while the water that seeps away removes salts. The khadin is, therefore, a land-use system, which prevents soil deterioration (Kolarkar *et al.*, 1983). This practice has a distinct advantage under saline groundwater condition, as rainwater is the only source of good quality water in such area.

c. Farm ponds

Farm ponds are very age old practice of harvesting runoff water in India. These are constructed by excavating soil from a low lying area of the field (Fig. 4.20). Following guidelines are to be followed in design and construction of farm ponds.

- High-storage efficiency (ratio of volume of water storage to excavation) can be achieved by locating the pond in a gully, depression, or on land having steep slopes. Whenever possible, use the raised inlet system to capture runoff water from the upstream. This design will considerably improve the storage efficiency of the structure.
- Seepage losses can be minimized by selecting the pond site having subsoils with low saturated hydraulic conductivity. As a rough guide, the silt and clay content of the least conducting soil layer is inversely linked with seepage losses. Therefore, it is best to select

the site having subsoil with higher clay and silt and less coarse sand. Also, reduce the pond wetted surface area in relation to water storage volume. This can be achieved by making the pond of a circular shape or close to circular shape.

- Evaporation losses can be minimized, if the ponds are made deeper but with acceptable storage efficiency to reduce water surface exposure and to use smaller land area under the pond.

(a) Check dam **(b) Farm pond**

Fig. 4.20: Common water harvesting structures

Water harvested in the pond could be used for providing supplemental irrigation to *kharif* crops during dry spells. Best returns are possible with cash crops or vegetables. Recent research by CRIDA resulted in design of low lift diesel run pumps (1.5 Hp) which are portable and can be used more profitably to lift water from ponds. This water can be conveyed either through sprinklers or drips (Fig. 4.21).

4.6.4 Development Efforts in Drought Management

4.6.4.1 *Short-term relief measures*

Drought relief measures are implemented as soon as the distress signals are visible. These measures are location-specific and require good cooperation and coordination among various departments. It requires continuous flow of information from village to administrative units and back. Some of the measures to reduce impact of drought implemented by Government of India through State Governments are described below.

Redgram

Okra

Portable pump sets of 1.5 hp petro-start diesel engines are suitable for 1 hectare farm size for sprinkler irrigation

Fig. 4.21: Pond water for life saving irrigations using portable pumpset

4.6.4.1.1 *Relief employment*

Central and State Governments initiate relief works to provide employment during drought. Most recently, the Government of India has started the National Rural Employment Guarantee Scheme (NREGS) in 2005 by providing 100 days of employment per year to adult members of family who were willing to undertake manual work for creation/ development of public assets at the statutory minimum wage. The Government is planning to increase days to 150 in drought years due to its success rate. The program also permits individual beneficiary works such as farm ponds, new wells/deepening of existing wells, construction of new water channels, rainwater harvesting structures, all of which contribute to drought proofing.

4.6.4.1.2 *Augmenting drinking water supply*

Water is the basic need for human and cattle population. Supply of drinking water during droughts is the responsibility of the State

Government. Measures taken by the Government for management of water resources includes (1) repair and augmentation of existing water supply schemes, (2) special measures and schemes for areas with acute drinking water scarcity, (3) construction of temporary piped water supply, (4) digging of new bore wells, and (5) supply of water through tankers and bullock carts.

4.6.4.1.3 *Ensuing food grains availability*

Supplying food grains through Public Distribution System (PDS) was initiated from 1965–66 which is jointly operated Central and State Governments. The PDS with a network of about 4.74 lakhs fair price shops in the country is one of the largest networks in the world. Since 1977, targetted PDS was introduced which follows a two tier subsidized pricing for people "Below Poverty Line (BPL)" and "Above Poverty Line (APL)". Community kitchens for certain segments of people such as old, disabled and women are run either by Govern-ment or through NGOs during extreme drought situations.

4.6.4.1.4 *Cattle camps and fodder supply*

To prevent the distress sale of cattle during drought years, the State Governments organize cattle camps in the affected regions and protect animals against starvation and diseases by transportation of fodder and feed from the areas where the situation is not alarming and need based vaccination. Government also encourages farmers in drought prone areas to undertake fodder cultivation on the banks of canals, tank beds or other areas under irrigation.

4.6.4.1.5 *Contingency crop planning*

Contingency crop planning is the most important short term strategy for drought management. The State Agricultural Universities and ICAR Institutes have formulated contingency plans for delayed monsoon, mid-season breaks and early withdrawal. These are available at state or agroclimatic zone level. In order to make them more useful and effective, ICAR, in collaboration with State Agricultural Universities and State Agricultural Departments are preparing district level contingency plans by pooling and analyzing all the relevant natural resources information of each district to cover all weather aberrations like droughts, floods, cyclones, heat wave, frost, etc. It is planned to prepare these plans for all 600 districts in the country by the year 2010–11. In addition to crops, livestock and fisheries are also covered in this program.

4.6.4.1.6 *Agro-advisories*

The India Meteorological Department (IMD) through its Integrated Agro-advisory Services are issuing weather-based advisories at the district level covering all 127 agroclimatic zones using the medium range forecast (3 to 4 days) and the crop conditions in collaboration with ICAR Research Institutes and the State Agricultural Universities. All the centres of All India Coordinated Research Project on Agrometeorology are also participating and the research results generated through this project are being used for formulating advisories. These are disseminated through various mass media and placed on the websites of ICAR, IMD and SAUs. A model contingency plan details are presented in Annexure 1.

Scientists of State Agricultural Universities also reach farmers through radio, television, newspapers and contact farmers. The economic benefits accrued due to agro-advisories can be seen in Table 4.56.

Table 4.56. Impact of Agro-advisories on farmer's income

Location	*Crop*	*Advisory given*	*Benefit to the farmer*
Jammu	Maize-Rice	Timing of irrigation	Rs. 750/ha
Trissur	Banana	Timing of irrigation	Rs. 1275/ha
Kovilpatti	Cotton	Pre-monsoon sowing	Rs. 1000–1500/ha
Bhubaneswar	Kharif Rice	Delaying pesticide spray	Rs. 850/ha
Ranga Reddy	Grapes	Spraying of fungicide	Rs. 31000/ha

4.6.4.1.7 *Use of ICT: Decision support system (DSS) for drought management*

A Decision Support System (DSS) is an integrated interactive computer based system consisting of analytical tools and information management capabilities designed to aid decision makers. A DSS typically consists of data bases, modeling tools and documentation of the decision making process. DSS could be used to provide timely information which will support the decision makers.

CRIDA has developed drought management plan for 1200 mandals and 22 districts of Andhra Pradesh. The mandals prone to drought were assessed and prioritized using bio-physical and socio-economic parameters. Drought severity indices were worked out for all the mandals. A DSS was developed with diverse modules dealing with assessment, mitigation and relief measures to reduce the time lag in collection, processing and transfer of data/information (Fig. 4.22). Forewarning for agricultural drought was attempted for the first time on real time basis. The inputs include daily rainfall, maximum and minimum temperature for deriving PET, crop coefficient, available water content (AWC) for a given soil type, maximum yield of crop, duration of the crop and sowing date. Information on groundwater, surface water and livestock management are also included in the DSS. Drought preparedness is much more cost-effective than expenditure on relief measures.

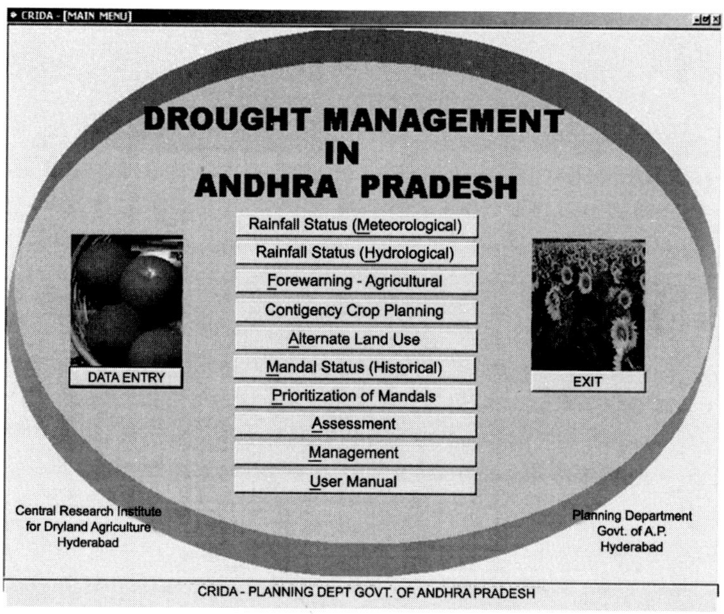

Fig. 4.22: Decision support system for drought management in Andhra Pradesh

4.6.4.1.8 *National agricultural drought assessment and monitoring system (N-ADAMS)*

An agricultural drought information and monitoring system has been launched in India using satellite data under the project 'National

Agricultural Drought Assessment and Monitoring System (NADAMS) in 1989 by National Remote Sensing Agency (NRSA). Under NADAMS, agricultural conditions are monitored at state level using daily observed coarse resolution (1.1 km) NOAA AVHRR data for the entire country and at sub district level for Andhra Pradesh and Karnataka using higher spatial resolution IRS WIFS data (188m). Fortnightly bulletins were issued at district level for 11 states in the country during 1989 to 1991. Since 1992, detailed monthly reports at the end of August for June, July and August and the monthly updates thereon for the period of September and October are prepared and sent to the concerned authorities for preparation of contingency plans.

4.6.4.2 *Medium-term strategies*

4.6.4.2.1 *Integrated watershed management programme (IWMP)*

The Government of India considers Watershed Program as the principal vehicle for drought proofing in rainfed areas. Till 2007 investments made by various agencies on watershed implementation are Rs. 194706 million and the area treated is about 56.5 m ha with an average investment of Rs. 3444 per hectare as per the assessment of Planning Commission.

In 1999–2000, NABARD has established a Watershed Development Fund (WDF) to enable states to access credit for treatment of large areas under watershed development (Government of India, 2000). National Rainfed Area Authority (NRAA) has recently formulated new common guidelines for Watershed Programs in 2008 where large area upto 5000 ha are to be covered which provides opportunities for resources conservation and drought proofing.

4.6.4.2.2 *Integrated water resources management*

Since water has an economic value, it is important that the competing uses for water must be reconciled in order to manage it sustainably. The Integrated Water Resources Management (IWRM) strategy of the Government of India aims to achieve this objective by considering water as a single resource unit regardless of different uses to which it is put to. IWRM aims to ensure the coordinated development and management of water, land and related resources by maximizing economic and social welfare without compromising the sustainability of the vital ecological systems.

4.6.4.2.3 *Improving water use efficiency*

Even after all the ultimate irrigation potential is developed in India, more than 50 percent of the cultivable land will still remain under rainfed cultivation. In addition to improving the performance of existing irrigation systems, higher efficiency in water use need to be ensured to meet increasing demands. ICAR and SAUs through their research efforts have evolved many improved irrigation methods, zero tillage, raised bed planting, breeding low water requiring varieties, etc. for enhancing the water use efficiency. However, micro-irrigation methods will be the key component for achieving water use efficiency.

4.7. MANAGEMENT OF DEGRADED LANDS THROUGH AGROFORESTRY

Agroforestry (AF) systems have an important role in natural resources management. These systems conserve soil, water, improve micro-climate and moderate the effects of extreme weather conditions, besides, providing economic products like timber, fruits, fodder and fuelwood (Singh, 2003). AF systems are classified into 3 sub-systems based on the nature of components, *viz.*, agrisilviculture, silvipasture and agrisilvopasture (Nair, 1985). Growing of trees with arable crops is termed as agrisilviculture while with forage crops is silvipasture and a system of crop + animals/pasture + trees is known as agrosilvipasture (Fig. 4.23). AF systems are planted both in arable and non-arable lands for control of soil and water erosion and as shelter belts/wind breaks for moderating micro-climate. These systems protect the farmers by way of insurance under extreme drought conditions when arable crops completely fail.

4.7.1 Agrisilvicultural Systems

In agrisilvicultural systems minimizing tree competition with annual crop is the key strategy for achieving satisfactory productivity. In fast growing trees like *L. leucocephala*, an acceptable harvest of annual crop was possible through canopy management by pollarding of trees

(Hocking and Rao, 1990). Competition can also be managed through hedgerow intercropping or alley cropping. In the semi-arid tropics (SAT), competition for soil moisture is the major factor causing reduction in crop yield in alley cropping (Rao *et al.*, 1991). Frequent pruning of hedgerows reduces the competition for light. In agrisilviculture systems, presence of a well-developed tree canopy and resultant shade makes light an important determining factor. Singh *et al.* (1989) observed sorghum and pearlmillet (C_4) to be the most shade tolerant crops among various cereals with yield averaging about 75% of pure crop in wide alleys across various sites and years. Rao *et al.* (1991) observed sorghum to be the most shade tolerant cereal crop compared to oilseed crops such as: sunflower, groundnut and castor bean. In a low rainfall year, the adverse impact of *leucaena* was found to be more severe on sunflower than on sorghum. Lal (1989) observed a 10% decline in grain yield of maize (C_4) in alley cropping on gross area basis compared to pure maize and up to 50% decrease in the yield of cowpea (C_3).

Fig. 4.23: Mango + Stylosanthes + sheep system

Leucaena leucocephala hedgerows continues to grow even during post-rainy period and yield 2 to 5 t ha^{-1} of green fodder when annual cropping is impossible under SAT (Ong, 1991). In terms of soil and rainwater conservation, agrisilvicultural system was superior to sole tobacco in a study in coastal Andhra Pradesh but poorer than pure pasture system (Osman and Srinivasulu, unpublished data). The cumulative rainfall for selected five runoff yielding events was 471 mm. The values of runoff and soil loss for pasture system were 108 mm and 2.2 t ha^{-1} compared to 144 mm and 3.0 t ha^{-1} for agroforestry system, respectively. Sole tobacco recorded the highest amount of runoff (241 mm) and soil loss (8.1 t ha^{-1}) indicating high susceptibility of arable cropping to natural resource losses. Plantation of *Leucaena leucocephala* reduced 40–80% runoff and 12.5t ha^{-1} soil loss in comparison to fallow plots on mild slopes up to 4% in areas receiving rainfall of up to 1000 mm (Narain *et al.*, 1998).

4.7.2 Silvipastoral Systems

Soil erosion is often more severe on pastures than on croplands. Degradation of the vegetation through overgrazing is a common feature, leading to erosion. Silvi-pastoral practices include scattered trees on pastures (systems with *Acacia* species), combination of plantation crops with pastures, live fences, fodder banks, windbreaks and shelterbelts and hedgerow intercropping on pastures. Planting trees alone will not restrict or reduce soil erosion, but other basic requirements of pasture management like restriction of livestock numbers and rotational grazing will have to be followed (Singh, 2003). Trees in silvipastoral systems supply protein-rich fodder during times when grass is absent or indigestible. By reducing grazing pressure, better vegetation cover could be obtained, thereby reducing erosion at the critical period.

Among various systems, the maximum potential for rehabilitation of degraded lands exists through silvipasture. India has about 120.72 m ha of degraded and wastelands (Anonymous, 2008) which can rehabilitated by AF systems. Some proven AF systems for different climatic zones in the country are listed in Table 4.57.

4.7.3. Hortipastoral Systems

In arid and semi-arid areas, inter-planting of grass strips with fruit trees will provide not only additional income but also reduce wind erosion. Care should be taken that first trees are planted one year a/

Table 4.57: Agroforestry practices for rehabilitation of degraded and wastelands

Agroforestry practice	*Environments in which it is applicable*	*Details*
Alley cropping or hedge row intercropping	Humid, sub humid and possibly semi-arid climate	An option to control soil erosion on gentle to moderate slopes and also in steep slopes up to 30%. The distance between alleys depends on the slope and need to be aligned on contour and maintained as hedge by repeated pruning. Suitable species are: *Leucaena leucocephala, Gliricidia maculata, Cassia iamea.*
Silvopastoral practices	Arid and Semi-arid areas	Trees and grasses can effectively control the soil erosion on mild to steep sloppy areas. Afforestation with suitable trees like *Acacia, Azadirachta, Prosopis, Dalbergia, Bambusa, Dendrocalamus* with grasses like *Cenchrus, Dicanthium, Borth-ricloa, Cynodon, Sehima,* etc. will help in im-proving the productivity of degraded lands
Trees and shrubs as bio-engineering measures	All zones	Species like *Gliricidia maculata, Pongamia pinnata, Lawsonia inermis, Jatropha curcas,* etc. can be taken up on bunds for stabilizing and strengthening them and also on field boundaries.
Wind breaks and shelter belts	Semi arid and arid zone	Proven potential to reduce wind erosion, stabilizes and dunes, provides supplementary products. *Acacia* spp., *Crotalaria burhia, Leptadenia pyrotechnica* and *Arva psuedotome-ntosa* are some of the trees used for such purpose in arid regions.
Woodlots	Arid, semi-arid and sub humid zones	The practice of growing trees in high densities with intercrops during the initial years for meeting the requirements of fuel wood, pulpwood, poles etc. Can be grown in highly degraded lands for arresting erosion and soil stabilization. Poplar plantations in north India, bamboo, leucaena and casuarina in coastal sands.

head of the establishment of grasses. A hortipastoral system of guava/ custard apple + *Stylosanthes hamata/Cenchrus ciliaris* was evaluated at CRIDA, Hyderabad, India. The growth of fruit trees 18 months after pasture establishment was found to be poor with grass than with legumes association. In terms of herbage yield, *Cenchrus* grass out yielded stylo legume but fruit plants suffered severely due to moisture stress (Singh and Osman, 1995).

Vegetative barriers with rows of trees, shrubs, perennial grasses, hedges, windbreaks and shelterbelts etc. on contours are cheap and effective barriers as compared to mechanical measures in reducing/ controlling erosion on mild slopes (Narain *et al.*, 1998). The barrier function is operative by reducing wind speed, checking runoff and suspended sediments. The cover function involves reducing raindrop impact and wind action on soil particles by increasing soil cover through litter and pruning. In black soils, leucaena keylines established at 0.25 m vertical interval recorded minimum runoff and gave maximum net monetary returns under pigeon pea+sorghum and pigeon pea+ soyabean system in sloping land at Parbhani (Jadhav *et al.*, 2006). At Kanpur, Leucaena based wider alley width (6.5 m) and intra row spacing of 0.5 m proved very effective in arresting erosion and contributed to higher sorghum yields (Haider *et al.*, 2005). Vegetative barriers influence redistribution of sediment in the catchments resulting in deposition of sediment on contours and forming terraces/bunds naturally. Live fencing with tree species such as *Acacia caesia, A. concinna, A. nilotica, Agave sisalana, Annona squamosa, Borassus flabellifer, Carissa carandus, Lawsonia inermis* and *Murraya koenigii* around fields and homesteads is an important indigenous practice in Eastern Ghats region of India which can control erosion (Chaudhury *et al.*, 2005).

Agroforestry systems provide an excellent alternative for resource management and income generation both arable and non-arable lands. In view of the additional benefit in terms of carbon sequestration and moderation of climate change, these systems need to be adopted more widely in future.

4.7.4 Crop Insurance Scheme

Crop insurance was introduced in India during *kharif* 1985, at the Comprehensive Crop Insurance Scheme (CCIS). One of the basic objectives of the scheme was "to provide a measure of financial support to farmers in the event of crop failure due to droughts, floods etc". But the scheme was voluntary and states were free to join the scheme. Only few food grain crops such as wheat, paddy, millets (including maize) and oilseed and pulses were covered in the scheme. Based on initial experiences, the scheme was revised as Experimental Crop insurance Scheme (ECIS). This was further improved and a Modified Comprehensive Crop Insurance Scheme (MCCIS) was introduced in *kharif*, 1999. In 2009–10, as many as 27 million farmers growing crops on 38 m ha were insured by Agricultural Insurance Company (AIC) alone.

4.7.5 Managing Seasonal Droughts

Drought management strategies rely mostly on efficient management of natural resources like soil and water. It is said that the preparation for a drought year is made in a good rainfall year. Some of the key *in situ* moisture conservation practices like inter-terrace land management, tillage and timely planting help in management of moderate droughts. Use of harvested rainwater for supplemental irrigation helps in management of mid-season and terminal droughts. Application of additional dose of nitrogen (10–20 kg N/ha) and additional interculture helta in managing mid-season droughts through better capitalisation of the moisture available during recovery (Subba Reddy, 2005). Indian farmers have evolved many traditional practices for coping with drought which have stood the test of time. These include *Khadins* in arid soils of Rajasthan, *Nadi* system in inceptisols of Bhilwara and percolation tanks in red soil regions of Andhra Pradesh and Karnataka. Research on water harvesting and recycling during last 30 years resulted in designing water harvesting structures of various sizes either on individual holdings on stand alone basis or as a part of the watershed management approach on a hydrological unit. Harvested water can be recycled to give a life saving irrigation to the kharif crop or pre-sowing irrigation for rabi crop.

4.7.6 Contingency Crop Planning

Change in crop planning and land use are important components of managing early season droughts while management practices become more relevant for coping with mid-season and terminal droughts. Research carried out under the AICRPDA in development of contingency plans for various climate zones and soil types which give a detailed account of crops, varieties and cropping systems can be grown in the event of delay in rainfall. These are listed in Table 4.58.

4.7.7 Drought Proofing Through Watershed Approach

Of late, integrated watershed management emerged as a key strategy of drought proofing. In this strategy, both resource conservation, crop management, alternate land uses like trees and pastures and livelihood diversification are integrated, thereby, minimizing risk and helping farmers to cope with the drought. The advantages of watershed

Table 4.58: Crop and contingency cropping plans for different drought types based up on climate and soil orders

	Crop Plan						
	June second fortnight to July first fortnight	*Second fortnight of July*	*First fortnight of August*	*Second fortnight of August*	*September first fortnight*	*September second fortnight to October first fortnight*	*October second fortnight to November first fortnight*
1. Chronic Drought							
1.1 Chronic drought in arid region in marginal soils with < 500 mm							
Hisar, Bhiwani, Sirsa, Mahendragarh, Gurgaon and part of Rohtak district of Haryana in marginal hot typic arid aridisols	Pearlmillet Cluster-bean Greengram	Short duration crops Pearlmillet (HHB-67) Greengram (Asha) Blackgram (T-9) Cowpea (Charodi) Clusterbeans (HG 365)	Nursery of Pearlmillet (HHB 67) nursery may be kept ready for trans-planting. Crops Fodder	Toria with supplemental irrigation Rains received after mid of August may be conserved for *rabi* sowing.	Clusterbean Cowpea	Chickpea Taramira Rapeseed	Mustard Taramira Use water in farm ponds, tanks and *tankas (Kund)* for drip and sprinkler irrigation for the horticulture, vegetables and high value crops.
1.2 Chronic drought in arid region in sub marginal soils with 500-750 mm							
Rajkot, Surendranagar, Jamnagar, parts of Junagarh, Bhavanagar and Amreli districts of Gujarat in hot arid vertisols	Sorghum, Pearlmillet, Groundnut, Castor, Cotton	Erect groundnut (GG-2, GG-5, GG-7) Sesame (G. Til-1, G. Til-2); Pearlmillet (GHB-235 GHB-316, GHB-558) Greengram (K-851, GM-4); Blackgram (T-9;) Pigeonpea (ICPL-87, GT-101)	Blackgram (T-9), Forage maize/sorg-hum (Gundri, GFS-5), Castor (GAUCH-1) Sesame (Purva-1)	Forage maize Sorghum (Gundri, GFS-5) Sesame (Purva-1)		Usually no rainfed crop is sown	

Table 4.58: (*Contd...*)

Table 4.58: (*Contd...*)

	Crop Plan						
	June second fortnight to July first fortnight	*Second fortnight of July*	*First fortnight of August*	*Second fortnight of August*	*September first fortnight*	*September second fortnight to October first fortnight*	*October second fortnight to November first fortnight*
1.3 Chronic drought in dry semi - arid region with 500-750 mm (late rainy season to post-rainy season)							
Maduari, Ramanthapuram and Tirunalveli districts of Tamil Nadu in hot moist semi arid Alfisols/ vertisols		Onset of monsoon is from September			Cotton Sorghum Pearlmillet Setaria Cotton Chillies Medium-duration pulses, Flowers, vegetables	First fortnight of Grain sorghum Third week of Chilies Fourth week *Panicum miliaceum* Chickpea Short-duration pulses Safflower Setaria Coriander Pearlmillet	First week *Karunganni* Cotton Sunflower and Sesamum sowings can be done during first week of November Senna No rainfed crop sowing beyond third week of Nov.
1.4 Chronic drought in dry semi – arid region with 500–750 mm (post-rainy season)							
Bijapur and Gulbarga districts, parts of Belgaum, Linguagur of	Greengram Blackgram Pigeonpea Pearlmillet	Groundnut (spreading) Hybrid pearlmillet Sunflower and	Cotton in middle of August. Early sowing of	In medium to deep black soils, on contour bunds	Early sowing is more beneficial. Cotton (Bhagya/	Sorghum, chickpea safflower (before	Fodder crops

Table 4.58: (*Contd...*)

Table 4.58: (*Contd...*)

	Crop Plan						
	June second fortnight to July first fortnight	***Second fortnight of July***	***First fortnight of August***	***Second fortnight of August***	***September first fortnight***	***September second fortnight to October first fortnight***	***October second fortnight to November first fortnight***
Raichur district of Karanataka and southern parts of Castor Maharashtra in hot semi-arid Vertisols		Setaria in *kharif* areas pure pigeon-pea/cowpea/horse-gram in light soils	cotton is adva-ntageous Grow herbaceum cottons in place of hirsutums Sunflower Pigeonpea, castor and setaria in light soils. Pigeonpea in medium to deep black soils	Castor Relay cotton in gro-undnut in medium black soils.	Laxmi _ before first fortnight of September If rains are not received seeding of sunflower, *rabi* sorghum, and chickpea with 1½ times the normal seed rate. sorghum, sunflower (after first dekad of September),	September) *Rabi* sorghum upto first fortnight of October *Rabi* sorghum + chickpea (2:1), *rabi* sorghum and chickpea as mixed crops chickpea and safflower (4:2 or 3:1).Wider row spacing or paired rows helps in moisture conservation especially during drought	
2. Ephemeral Drought							
2.1 Ephemeral drought in wet semi - arid region with 500–750 mm (rainy season)							
Bhilwara, Tonk, Dungarpur, Ajmer	Pigeonpea Cowpea	Sesame (RT-46) Greengram (K-851	Sesame (RT-125) Greengram	Sorghum (Fodder) (Raj	Greengram Blackgram	Greengram Blackgram	Chickpea Safflower

Table 4.58: (*Contd...*)

Table 4.58: (*Contd...*)

	Crop Plan						
	June second fortnight to July first fortnight	*Second fortnight of July*	*First fortnight of August*	*Second fortnight of August*	*September first fortnight*	*September second fortnight to October first fortnight*	*October second fortnight to November first fortnight*
districts and parts of Bundi, Chittaurgarh, Rajasamand in hot dry semi-arid Inceptisols/ Aridisols		& RMG-62); Sorghum (Fodder); Cowpea (Fodder) (Raj Chari-1 & 2, C-152) No sowing of cereals only short duration pulses and oil seeds or fodder crops should be sown. Soil mulching and interculture to conserve soil moisture is beneficial.	(RMG-62) Sorghum (fodder) (Raj Chari-1) (single cut) Soil water conserv-ation measures for *in situ* management & runoff harvest for recycling during later part of crop season.	Chari -1) Toria (TL-15) Tarm-aira (T-27) Rain received after first fortnight of August should be conserved for early *rabi* seeding of toria/taramira during first week of September. Any heavy downpour occurs, harvest the water for pre-sowing supplemental irrigation to *rabi* crops.	Cowpea	Cowpea	Mustard Barley

Table 4.58: (*Contd...*)

Table 4.58: (*Contd...*)

	Crop Plan						
	June second fortnight to July first fortnight	*Second fortnight of July*	*First fortnight of August*	*Second fortnight of August*	*September first fortnight*	*September second fortnight to October first fortnight*	*October second fortnight to November first fortnight*
3.Apparent Drought							
3.1 Apparent drought in dry sub humid region with 1000–1500mm (rainy season)							
Uplands and medium lands of Balasore, Cuttack, Puri and Ganjam districts of Orissa in hot moist sub-humid inceptisols	Pigeonpea Mesta Maize Groundnut Fingermillet Rice, Sorghum Cowpea, Blackgram, Greengram	***Upland*** Blackgram Setaria (Pant -30); Greengram (PDM-54/K 851);Sesame (Uma or local); Early Pigeonpea (UPAS 120/ICPL-87)Planting of short duration vegetables as radish (Pusa Chetki), okra, cowpea (SEB-2/ SEB 1) and clusterbeans.	***Upland*** Sowing of niger, black-gram, sesame, greengram, Planting of vegetables as radish, beans, cowpea, Early Pigeonpea (ICPL -87/UPAS-120)/ WMedium and shallow subme-rged lowland Direct line sowing of extra early rice (Heera, Vandana, Kalinga-III, Z HU 11-26, Rudra, Sankar and Jaldi-5).	***Upland*** Horsegram; Sesamum Niger; Cowpea	Medium land Making land prepar-ation for sowing pre-*rabi* crops as Mustard / Greengram / early pigeon-pea, which can be sown in the month of September Pigeonpea (short duration), Horsegram	Medium and low land conditions: Direct sown rice, transplanted rice Horsegram	–

approach in stabilizing production and income during low rainfall years has been widely documented.

The Government of India has been implementing special programmes in drought prone districts like DPAP and DDP for several years for enabling communities cope with droughts. In drought prone regions, implementation of natural resource management (NRM) technologies on watershed basis is more important which need proper institutional and policy support and hence the solution for such problems does not lie in research. Since many of the NRM technologies like community based SWC measures are labour and energy intensive, institutions such as custom hiring centres and policies like linking rural employment guarantee programmes with watershed development activities would help faster upscaling of technologies. A national land use policy for rainfed regions where in the development of drought prone regions as integral part is needed for developmental funding with special assistance to natural calamities like drought, with adequate institutional support for input supply, credit, value addition, marketing and price support. These efforts are likely to address drought in a more holistic manner. Despite this, drought management continues to pose major challenges for the Government in view of its unpredictability and wider socio-economic effects.

Other financial and policy instruments like drought prediction, development of early warming systems, strengthening agro-advisories on contingent planning and introduction of rainfall based insurance are likely to be more important in future and will greatly supplement drought management efforts through natural resource management.

4.8 CONSERVATION AGRICULTURE

There is a growing concern on enhancing farm productivity with increasing problems of land degradation, soil fertility decline and stress on water resources. Conservation agriculture (CA) has emerged as an effective alternative to sustain productivity along with resource conservation. CA is a generic term encompassing a number of crop and soil management practices that conserve soil, save water, labour, fuel and other inputs. CA either increases or maintains crop yields, while making effective use of rainfall and stored water. Research during the last decade in India and other Asian countries clearly brought out the advantages of CA over conventional farming besides conservation of soil.

4.8.1 CA in Irrigated Production System

No-till farming: Trials on rice-wheat cropping system in Indo-Gangetic plains have brought out the advantages of resource conservation technologies (RCTS) like no till planting, raised beds, direct seeded rice, laser land leveling, and crop residue management over traditional tillage and planting methods (Sharma, 2006). Traditionally, the crop establishment includes repeated plowing, cultivating, planking and pulverizing the top soil. Repeated tillage requires more time, energy, escalate costs, reduce profits and needs more water (Gupta *et al.*, 2003). Wheat can be planted without tillage in rice-wheat cropping system to address these issues. Several on station and on-farm trials have clearly brought out the advantages in cost and saving on resources with this system (Malik *et al.*, 2005). Its success has encouraged farmers adopt the double no-till practice in the IGP areas. No-till technology is scale neutral and can be practiced by all farmers. Timely planting of crops with no-till has many benefits for small farmers; (a) lowers production costs, (b) improves productivity, (c) saves irrigation, water and crop residues from burning and (d) reduces the acreage of rice-fallows.

Direct seeded rice: Rice-wheat rotation is the predomi-nant cropping system in IGP wherein rice is the major consumer of water. Traditionally rice is transplanted after puddling and wheat is sown after pulverizing the soil. Direct seeded rice (DSR), which avoids puddling and drudgery of transplanting the young rice seedlings, provides an option to resolve the edaphic conflict and enhance the sustainability of rice-wheat cropping system. The results of the trials under PDCSR-IRRI-RWC collaborative project (Table 4.59) revealed that with DSR technology, similar rice yields can be produced with less water. The DSR followed by zero till wheat and other upland crops have shown high potential as components of CA in heartland of rice-wheat production system.

Raised bed planting: During the last decade the practice of furrow irrigation raised bed (FIRB) planting has emerged as a successful intervention in IGP. Based on several on-station and on-arm trials, it was proved that yields of wheat planted on beds were similar to those

Table 4.59: Effect of tillage and crop establishment methods on water productivity and irrigation water savings in rice crop in IGP

Crop establishment	*Grain yield (t/ha)*	*Total no. of irrigation*	*Total irrigation water applied (mm)*	*Water productivity (kg grains m^{-3} water)*	*Water saving over conventional*
Puddled TPR	7.6	26	1872	0.41	–
CTDSR	7.6	27	1755	0.43	6.25
LSD (P = 0.05)	NS	–	–	–	–

Source: Jat *et al.* (2006)

planted with conventional till but with reduced irrigation amounts or time. Some of the main advantages are opportunities for relay cropping *e.g.* wheat-sugarcane, mechanical cultivation of weeds, reduced irrigation time, and improved yields on fine textured soils prone to water logging. The economic analysis of a maize-wheat system revealed that net profit (US\$ ha^{-1}) was more in no till (340 and 463) followed by FIRB (332 and 441) and least in conventional till (144 and 337), respectively for maize and wheat (Table 4.60). The water productivity was markedly higher with FIRB than other systems.

Table 4.60: Effect of no till, flat and raised bed planting system on crop yield, water productivity and profitability in maize-wheat system

Source	*Grain yield (tones ha^{-1})*			*Water productivity (kg grain m^{-3} water)*		*Net profit (US\$ kha^{-1})*		
	Maize	*Wheat*	*System*	*Maize*	*Wheat*	*Maize*	*Wheat*	*System*
No till	4.75	5.56	10.31	1.74	1.89	340	463	803
FIRB	5.66	5.18	10.84	2.79	1.98	332	441	773
Conventional till	4.39	5.12	9.51	1.36	1.38	144	337	481
Average	5.03	5.15	10.17	2.08	1.68	238	389	627

***Crop residue management*:** Decline in soil organic matter (SOM) due to limited return of organic biomass to soil owing to residue burning is a key factor for unsustainability in intensively cultivated irrigated production system of IGP. Burning crop residues due to lack of efficient and user-friendly technologies for *in situ* recycling (Jat *et al.*, 2004) not only leads to loss of considerable amount of N, P, K and S, but also contributes to emission of greenhouse gasses and evaporative losses of water from the soil. Surface managed crop residues with new generation drills (happy seeder, roto-disc drill, double disc drill, punch planter) in no till systems will regulate the soil water nutrient interactions, soil and plant canopy temperatures and saves on irrigation water with higher crop yields. Crop residues retained on the surface have improved the crop yields on an average by 247 kg ha^{-1} by reducing moisture loss and improving soil health (RWC-CIMMYT, 2006).

***Crop diversification*:** Legumes in the cropping systems of IGP have declined overtime leading to more unsustainability. Recent advances in RCTs have raised hopes of greater use of legumes in cereal-dominated systems of IGP, to increase crop diversification and contribute to system sustainability. The diversified agro-climatic, biophysical and socio-economic conditions of IGP have wide scope for diversification of rice-wheat cropping system. This diversification became feasible with the development of new generation multi-crop bed planters. Availability of extra short duration varieties (pigeonpea; ICPL-88039) and FIRB planting technique opened up tremendous opportunities for increasing the water productivity and economic returns for farmers with same resources through crop diversification (Table 4.61).

Table 4.61: Effect of crop establishment methods on crop and water productivity and profitability of rice and pigeonpea

	Yield (tones ha^{-1})		*Water productivity (kg grain m^{-3} water)*		*Net profit (US\$ ha^{-1})*	
	Rice	*Pigeon pea*	*Rice*	*Pigeon pea*	*Rice*	*Pigeon pea*
Flat bed	5.60	2.2	0.70	1.47	262	644
Raised bed	5.80	2.6	0.85	2.31	318	811
Average	5.70	2.4	0.78	1.89	290	728

Source: RWC-CIMMYT, 2005

4.8.2 CA in Rainfed Agriculture

So far not much success has been achieved with CA in rainfed agriculture. In arid (< 500 mm rainfall) region, low tillage was found comparable to conventional tillage and weed problem was manageable. In semi arid region (500–1000 mm), conventional tillage was found superior to reduced tillage where successful crop production depends on rainfall infiltration and holding soil moisture in the profile. Weed problem is ephemeral depending upon the seasonal rainfall distribution in sub-humid (< 1000 mm) regions, but there is a possibility of reducing the intensity of conventional tillage by using herbicide (AICRPDA, 1999–2003; Ramakrishna *et al.*, 2005).

A long-term study comprising of tillage (conventional (CT) and reduced (RT)) and conjunctive nutrient use (fertilizers and low cost farm-based organics) treatments was conducted in SAT Alfisol (Typic Haplustalf) at Hyderabad, India, under sorghum and mung bean cropping system. Tillage and nutrient management treatments and their interactions significantly improved the grain yields and sustainability yield index (SYI) of the crops studied. Effect of tillage on sorghum grain yield was found significant only in five out of seven years. Analysis of data upto 8th year of the study revealed that conventional tillage maintained 12.8% higher sorghum grain yield and 11.2% higher mung bean yield when compared to RT. In the ninth and tenth years, reduced tillage was almost equal to conventional tillage in terms of sorghum and mung bean yields (Sharma *et al.*, 2008). This raised the hope of getting future success in practicing reduced tillage in rainfed Alfisols which are shallow, with compacted sub-soil horizon (murrum layer) and susceptible to hard setting. The highest average sorghum grain yield was recorded with 2 t Gliricidia loppings + 20 kg N thru urea (1895 kg ha^{-1}) under conventional tillage and with 4 t compost + 20 kg N through urea (1580 kg ha^{-1}) under reduced tillage. In case of mung bean, the highest grain yields under conventional tilled plots was observed in 2 t compost + 10 kg N through urea (959 kg ha^{-1}), while under reduced tillage plots, it was under 2 t compost + 1 t gliricidia loppings (836 kg ha^{-1}). In another long-term experiment in sorghum-castor rotation conducted for a period of eight years, conventional tillage proved superior to minimum tillage in terms of yields of both sorghum and castor. The interactive effect of tillage and N on crop yield was significant (Sharma *et al.*, 2005). Lal (1997) reported that to derive effective advantages of reduced tillage in semiarid tropics, it is essential to maintain crop residue on the surface as land cover (Lal, 1997) which is big challenge in India due to competing demands as fodder for livestock. However, there are many opportunities for CA in rain fed crops where crop residues do not have attenuation uses such as maize, cotton pigeon pea, earths which are other wife burned (Srinivasaraw *et al.*, 2012c).

Rainfed areas have heterogenous soils with different constraints like sealing, cracking and hard setting. Ensuring good seed germination and crop stand establishment are major challenges to be addressed with reduced tillage and crop residue management. Rainfall being the only source of water, interactions between rainwater conservation and reduced tillage are to be studied in an integrated manner.

4.9 COPING WITH CLIMATE CHANGE

Climate change impacts on agriculture are being witnessed all over the world, but developing countries like India are more vulnerable in view of the fact that majority of the population depend on agriculture, leading to excessive pressure on natural resources and poor coping mechanisms. Important observations on climate change in India are : increase in temperature, particularly during post monsoon and winter, increased frequency of floods during the monsoon, decrease in winter precipitation, lower number of rainy days, and faster retreat of Himalayan glaciers. The likely impacts are reduction in yields of crops like wheat, rice, shifts in cropping zones, sharp changes in water resources availability overtime and space decline in livestock productivity and migration of fish. Improved technologies and new policy initiatives are needed to enable farmers cope with climate change impacts.

4.9.1 Climate Change Impacts on Major Crops in Indo-Gangetic Plains, Hilly Regions and other States of Central and Eastern India

The Indo-Gangetic Plains encompass large areas of Indian Sub-Continent is a very fertile belt and is a major contributing to the 'food bowl' of India. In India, the belt spreads across regions of Bihar, Haryana, Punjab, Rajasthan, Uttar Pradesh, West Bengal, Punjab, Madhya Pradesh, Jharkhand, and Jammu & Kashmir states.

The benefits of Green revolution were significantly realized in most of Indo Gangetic Plain (IGP) areas. However, in the recent years, the yields realized have not been able to sustain the pressure of growing food demands. The projected predictions indicate that South Asia will be deficient in cereal production by about 22 Mt by 2030. Rice and wheat are the major cereal crops of this region and grown in an area of 21Mha of land with a significant production and productivity.

4.9.1.1 *Possible impacts of climate change on IGP*

This region is an environmentally sensitive, socially significant and economically strategic domain of India where landscape, hydrology, and fertility are threatened by climate warming and anthropogenic pressure. One of the climate change impacts is uneven water distribution during and over seasons. For instance, increased water supply will result in erosion and sedimentation in piedmont zones and river lowlands. On the other hand, decreased water supply, might lead to salinization, drought desertification, diminished in-flow to rivers and drying-up of aquifers in upland regions. These impacts will have a significant bearing on farm productivity. To assess vulnerability to environmental changes, efforts should be initiated to understand the geological past and model the future for which a high resolution history of climate variations and their impact on agriculture is crucial.

Preliminary findings suggest that rise in atmospheric temperature reduced the biomass and yields of various crops, *viz.*, rice, wheat and pulses. On the contrary, the elevated CO_2 levels enhanced yield and yield-related attributes of various crops. The traditional model indicates that evaporation and humidity will increase due to warming, which in turn influence the precipitation. Some recent studies indicate that global warming could increase the overall precipitation by about 20 percent. Hence, rapid hazards like flood, erosion and deposition will possibly accompany. These impacts can change boundary conditions of landforms, land-water and vegetation.

Major research programs have been initiated across the country by Indian Council of Agricultural Research to assess vulnerability of Indian Agriculture to climate change and also develop suitable adaptation and mitigation strategies. The umbrella research program "Network Project on Climate Change" (NPCC) spreads across 23 centres covering all commodities that contribute to the food security of India. The All India Coordinated Research Project on Agromet-eorology (AICRPAM) that was initiated by ICAR in 1983 also focuses on agroclimatic characterization, crop weather relationships, crop growth modeling, Agromet Advisories and weather influence on pests and diseases. The state-wise salient findings of these programs is summarized below.

4.9.1.1.1 *Punjab*

The magnitude of response of crop to elevated CO_2 and temperature varies with type of crop, region and climate change scenarios. At Punjab, simulated yield of both rice and wheat was decreased by 15–17 percent with a temperature rise of 2°C. But beyond this range the decrease of grain yield was very high compared to rice. Rise of temperature during the reproductive stage of wheat, from mid-February to mid-March affects the wheat yield realization very adversely. The temperature increase mostly affected the early (October) sown crop during the 4th week of January, February and up to 1st fortnight of March; the normal (November) sown crop during February and March; late (4th week of November) sown crop during March; and very late sown crop during March and 1st week of April.

4.9.1.1.2 *Haryana*

Like Punjab, at Haryana rice and wheat are grown in irrigated double cropping rotation with current total production of 11 million tonnes. The state is witnessing increase in rainfall receipt during the monsoon period in last 32 years (1970–2001), where as temperature is showing an increasing trend on annual basis. Incidence of intense foggy events that curtails the photo synthetically active radiation, intense spell of high temperature during grain filling and ripening stages are attributed to poor yield realization and quality deterioration of wheat in Haryana. Grain yield reduction of up to 20–30 percent is likely in areas where the crop can not be irrigated frequently during February and March.

4.9.1.1.3 *Uttar Pradesh*

Uttar Pradesh is also one of the most productive states in the IGP with highly fluctuating yield levels of rice and wheat due to aberrant weather. In Eastern Uttar Pradesh, the magnitude of losses during *kharif* season in rice crop due to increase in maximum temperature was 2 to 2.5 q/ha. Terminal droughts are the other sensitive factor for rice yield reduction. The model estimations showed that terminal drought reduced the rice yields by 44% compared to early season drought (23%). In wheat, under normal sown conditions (25 Nov) the crop experiences about 15 percent yield reduction due to above normal temperature of about 3–5°C during flowering phase. In the central plain zone of UP, the vegetative stage of wheat crop has significant negative correlation with bright sunshine hours.

4.9.1.1.4 *West Bengal*

Inundation of coastal lands by rising sea water, rapid degradation of red lateritic soils which are already poor in nutrient and water retention are the major manifestations of climate change in West Bengal affecting its mainly rice-based agriculture very adversely. Different climate change scenarios projected warming of Gangetic West Bengal and its surrounding regions by 0.4 ± 0.2°C. On rainfall scenarios, an increase in rainfall is estimated in all the seasons except in the pre-monsoon period.

Simulation study using WOFOST crop model in New alluvial zone of West Bengal suggested decrease in yield of *kharif* rice due to increase in temperature by 1°C, both maximum and minimum, is to the tune of 300 kg/ha. If the temperature rises by 2°C, the potential yield will further reduce up to 800 kg/ha.

4.9.1.1.5 *Madhya Pradesh*

Chickpea and soybean are major crops grown in Madhya Pradesh. A significant increase in crop season temperatures ranging from 1.8 to 5.6°C has been observed at Madhya Pradesh. On an average, an increase in rainfall by 10–25 percent is projected from different climate change scenarios. Large spatial variability in the quantum of change in temperature and rainfall across major soybean and groundnut growing regions is also anticipated.

In Madhya Pradesh, the average simulated rainfed yields of soybean (2144) and groundnut (2473) are more than double of national average, but the existing potential is not being exploited properly. Simulation study using Infocrop model has suggested that in case of soybean, yields are projected to increase (8–13%) under future climate scenarios (combined effect of increased temperature, CO_2 and change in rainfall). In case of groundnut except for A1B 2071–2100 scenario (–5.2%), yields are projected to increase (5–7%) under future climate scenarios. Across locations, the rainfed yield of soybean and groundnut shows significant positive association with crop season rainfall, but association with temperature was poor/non-significant. Thus, it is suggested that under rainfed conditions, the availability of water will remain a major factor determining the overall productivity of soybean and groundnut in India.

4.9.1.2 *Other States*

4.9.1.2.1 *Chhattisgarh*

Chhattisgarh state has three agroclimatic zones, *viz.*, Chhattisgarh Plains, Baster Plateau and Northern Hills. The climates of the zones are different from each other. Some part of the Northern Hill zone exhibit temperate climate whereas, southern part of Baster zone resemble tropical type of climate.

Spectacular reduction in annual rainfall amount between 1906 and 2000 at Mahasamund region of Chhattisgarh exhibits transition of climate from moist sub-humid to semi-arid conditions. That led to onset of desertification process along with drastic reduction in rice yield in that region. To counter the adverse impacts of desertification, crop diversification has been given importance in that region.

In Northern Hill zone, prominent reduction in maximum temperature during March has been taken place over the years and that prompted farmers to take up two crops of potato after the harvest of early/short duration rice.

Rice, both rainfed and irrigated, is the main *kharif* crop in Chhattisgarh. The effect of increased temperature during the reproductive stage was found detrimental for rice in both the conditions. A simulation study using CERES-Rice revealed that increase of maximum temperature during the reproductive stage by 1 and 2°C above normal can cause rice yield reduction by 3–4 and 5–6 q/ha, respectively. Further the overall crop duration would also decrease affecting net productivity. If the rice reproductive duration is decreased by 7 days, that would result in decrease of rice potential yield by 10 q/ha. If water stress is accompanied by increased temperature, the yield reduction further goes up to 15–18 q/ha.

4.9.1.2.2 *Orissa*

Orissa is actually located in East Coast of India and rice is the main crop grown across the state. While the coastal districts of Orissa witnessed increase in mean temperature during 1951 to 2006, the interior districts recorded slight fall in the same parameter during the above period. Amount of rainfall receipt showed a stable trend but spatial variation of rainfall was observed, though not significant statistically.

Dry-matter as well as grain yield in rice is expected to reduce at Orissa by 2020 if temperature rise occurs above 30°C, which is considered as critical temperature. The reduction in yield may be attributed to high rate of spikelet sterility, decreased growth duration particularly at grain filling period and high respiratory loss without change in rate of photosynthesis.

Delayed onset of monsoon and decreased rainfall in June led to more crop loss in rice under broadcast system than the transplanted rice and already forced the farmers to adopt transplanting for more assured yield. Water requirement will rise tremendously owing to high rate of evaporation loss. Increased infestation of rice crop by swarming caterpillar and hispa, and incidence of bacterial leaf blight (BLB) disease have already been reported from many parts of Orissa every year. Besides, increased frequency of heavy rainfall events augment flood risk, which would damage standing crops, increase soil erosion and make productive lands waterlogged.

4.9.1.2.3 *Some suggested adaptive measures*

The output of some other studies so far indicated that a marginal 1°C increase in atmospheric temperature along with increase in CO_2 concentration would cause very minimal reduction in wheat production of India if simple adaptation strategies like adjustment of planting date and varieties are adopted uniformly. But in absence of any adaptive mechanism the yield loss in wheat go up to 6 million tonnes. Study on wheat and rice suggested that high temperature around flowering reduced fertility of pollen grains as well as pollen germination on stigma. These effects are more pronounced in *Basmati* rice as well as *Durum* wheat cultivars. A positive finding of the study was that the *Aestivum* wheat cultivars are more or less tolerant to such adverse affects. But differential impact of increasing temperature is observed with respect to grain quality of wheat where it is found that *Aestivum* wheat cultivars are more prone to reduced grain quality due to increasing temperature during the fruit setting stage than *Durum* cultivars.

4.9.1.3 *Hilly Regions*

4.9.1.3.1 *Uttarakhand*

Agriculture is taken up in the valley and tarai areas of this region. The analysis of rainfall data indicates that there is a reduction of 23.8 percent in rainfall and 17.6 percent in rainy days in the mid Himalayan region. Higher temperatures at reproductive and maturity stages reduced the wheat yields. Erratic distribution of rainfall during *rabi* season due to missing western disturbance activity over the region adversely affecting the wheat yields of the region. Production of apple is vulnerable now a days due to non availability of optimum chilling time of more than 1200 hours in association with frequency of snowfall *i.e.* 4 times during the season.

4.9.1.3.2 *Himachal Pradesh*

Significant decreasing trend in rainfall was observed at Chamba and Solan and increasing trend at Una. Delayed tendency of onset during past 3 decades maximum temperature showed significant increasing trend during October, November and December, whereas minimum temperature showed significant decreasing trend during July to December. Decreasing trend in rainfall during July to October was also noticed. Analysis of recent weather data in Himachal Pradesh indicated that the maximum temperature is showing an increasing trend during November to April. This has resulted in a possible shift of apple belt upwards (and thus limited fulfillment of chilling hour's requirements of the crop) and increasing area of apple in higher elevations. The new areas of apple cultivation have appeared in Lahaul and Spitti and upper reaches of Kinnaur district of Himachal Pradesh. Trend predicted apple productivity in Kullu, Shimla as well as overall average productivity of the state has shown declining trend. Average state productivity in 1980–81 was 7.06 tonnes ha^{-1} that decreased to 4.65 tonnes ha^{-1} in 2004–2005.

4.9.1.3.3 Some suggested agronomic manipulations for the hill eco-systems

- The suggested horticultural crops in the vacated apple growing regions are pears, apricots, plums and kiwi fruits to match to the increased temperature in the mid Himalayan regions.
- In North West Himalayas off-season vegetable cultivation is highly remunerative enterprise. Produce fetches highly remunerative prices to growers during off-season.
- The sowings of wheat can be delayed upto first week of December without much reduction in yield as the region remains dry during post rainy season.
- Some of the heat tolerant apple varieties *viz.* Gale-gala and red-fungi can be grown.
- Mixed cropping system which has been developed over a wide range of variation in adaphic, topographic and climatic conditions over centuries should be given priority which will ensure some income even if some crop fails due to biotic and abiotic stresses.

4.9.2 Adaptation and Mitigation Technologies

While research on mitigation and adaptation of agriculture to climate change is at a nascent stage in India, based on knowledge already generated, some options for crop production under climate change induced effects like droughts, high temperatures, floods and rising sea level can be suggested. Adaptation to climate change in arid and semi-arid regions depends on the cost of adaptive measures, existence of appropriate institutions, access to technology, and biophysical constraints such as land and water resource availability, soil characteristics, genetic diversity for crop breeding (*e.g.*, development of heat-resistant cultivars), and topography.

The successful tackling of the direct effects of climate change on grain crops *viz.*, reduction in duration, embryo abortion, spikelet sterility, effects on grain number and grain size, anthesis interval etc, can be done by use of conventional breeding and molecular/ mutation breeding by the use of biotechnological tools including marker assisted selection, whole genome expression analysis and its subsequent elucidation and gene finding by bioinformatics. The indirect effects such as decline in water resources, increased pests and disease incidence, loss of soil organic carbon should be addressed by efficient use of water, carbon, energy; Integrated Pest Management (IPM) and conservation agriculture (CA). The adaptation and mitigating technologies may be crop based or resource management based (Singh, 2008).

***4.9.2.1 Crop based technologies*:** These are mainly centered on promoting the cultivation of crops and varieties that fit into new cropping systems and seasons, development of varieties with changed duration that can overwinter the transient effects of change, release of varieties for high temperature, drought and submergence tolerance; evolving varieties which respond positively in terms of growth and yield to high CO_2. In addition, varieties with high fertilizer and radiation use efficiency and also novel crops and varieties that can tolerate coastal salinity and salt water inundation are needed.

Agricultural biodiversity and crop germplasm exploration for favorable traits is an important area that needs to be tapped to the fullest extent. Seeds, plants and plant parts exhibiting tolerance to temperature, water and other atmospheric stresses caused by climate change needs to collected and conserved to aid crop breeding research. A thorough revisit and reevaluation of all the wild relatives, land races, extant varieties, modern varieties and breeding stocks could help in unraveling previously unknown or ignored traits than could prove more than useful in the present scenario. Genetic resources could well prove to be the most important cost effective basic raw material which will allow agriculture to adapt to climate change in India. Hence the key activities in the area of genetic resource conversation are: regenerate threatened collections, ensure safety, duplicate collections, upgrade key genebanks, develop global accession level information systems, and promote improved methods to conserve difficult crops (*e.g.* banana, yam, coconut). The genetic resources, especially land races from areas where past climates mimicked the projected future climates could serve as the starting material for building the genes for tolerance, maturity and yield features.

Strategies for genetic enhancement of heat tolerant genotypes specially in pulses by identifying and validating markers for high temperature tolerance coupled with yield potential is one of the key technological advances that can have a significant impact on combating climate change induced food insecurity in India. An additional strategy is to take advantage of faster growth under higher temperatures. The new varieties, especially of the rabi cropping season should have characteristics of early flowering (photo- and temperature-insensitivity, but development-related onset of flowering) and early maturity and high yielding.

Improved and novel agronomic and crop production practices like adjustment of planting dates to minimize the effect of temperature increase-induced spikelet sterility can be used to reduce yield instability, by avoiding the flowering period to coincide with the hottest period. Adaptation measures to reduce the negative effects of increased climatic variability as normally experienced in arid and semi-arid tropics may include changing the cropping calendar to take advantage of the wet period and to avoid extreme weather events during the growing season. Crop varieties that are resistant to lodging may withstand strong winds during the sensitive stage of crop growth. In addition improved crop management through conservation agriculture, crop rotations and intercropping, integrated pest management, supplemented with agro forestry and afforestation schemes will be an important component in strategic adaptation to climate change.

Intercropping is an efficient strategy that can be followed with desirable outcome in the present climate change scenario. Grain-legume intercrops have many potential benefits such as stable yields, better use of resources, weeds, pest and disease reductions, increased protein content of cereals, reduced N leaching as compared to sole cropping systems. Promotion of scientific agro-forestry forms yet another key component in adaptation and mitigation. Agroforestry systems buffer farmers against climate variability, and reduce atmospheric loads of greenhouse gases through carbon sequestration.

***4.9.2.2 Resource management technologies*:** The key resource conservation based technologies in combating climate change in India are; *in situ* moisture conservation, rainwater harvesting and recycling, efficient use of irrigation water, energy efficiency in agriculture and use of poor quality irrigation water. Strategies that have to be adopted to achieve the goals are: characterization of bio-physical and socio-economic resources utilizing GIS and remote sensing; integrated watershed development; improved rainwater use efficiency through better storage, delivery and application; contingency crop planning to minimize loss of production during drought/flood years.

Some potential resource management technologies which help in coping with climate change are zero-tillage (ZT), and bed planting. In rice-wheat cropping system in Indo-Gangetic plains, ZT and bed planting not only saved on fuel, water and energy but also helps in building soil carbon. Likewise for the coastal salinity in Tamil Nadu and Andhra Pradesh, Doruvu Kottai technology has been developed which allows farmers use normal water on a sustainable basis where sea water intrusion is a problem. System of rice intensification (SRI) is another practice which saves on resources and indirectly helps coping with climate change. Besides site specific nutrient management for a large number of crops, fertilizer management in paddy based cropping system also help in decreasing the green house gas emissions without affecting the yields.

Soil carbon sequestration is yet another strategy towards mitigation of climate change. Although, tropical regions have limitation of sequestering carbon in soil due to high temperatures, adoption of appropriate management practices helps in sequestering reasonable quantities of carbon in some cropping systems particularly in high rainfall regions (Srinivasarao *et al.*, 2009c). The potential of cropping systems can be divided in to that of soil carbon sequestration and sequestration in to vegetation. Tree based systems can sequester substantial quantities of carbon in to biomass in a short period. Total potential of soil C sequestration in India is 39 to 49 Tg year^{-1} (Lal, 2004). This is inclusive of the potential of the restoration of degraded soils and ecosystems which is estimated at 7 to 10 TgC year^{-1} (Table 4.62). The potential of adoption of recommended package of practices on agricultural soils 6 to 7 Tg year^{-1}. In addition, there is also a potential of soil inorganic carbon sequestration estimated at 21.8 to 25.6 TgC year^{-1}. Long term manurial trials conducted in arid regions of Andhra Pradesh (at Anantapur) under rainfed conditions indicate that the rate of carbon sequestration in groundnut production system varied from 0.08 to 0.45 t ha^{-1} year^{-1} with different nutrient management systems (Srinivasarao *et al.*, 2009b). Under semi arid conditions in Alfisol region of Karnataka, the rate of carbon sequestration was 0.04 to 0.38 t ha^{-1}

Table 4.62: Soil organic carbon sequestration potential through restoration of degraded soils

Degradation process	*Area (Mha)*	*SOC sequestration rate (kg/ha/y)*	*Total SOC sequestration potential (Tg C/y)*
Water erosion	32.8	80–120	2.62–3.94
Wind erosion	10.8	40–60	0.43–0.65
Soil fertility decline	29.4	120–150	3.53–4.41
Water logging	3.1	40–60	0.12–0.19
Salinization	4.1	120–150	0.49–0.62
Lowering of water table	0.2	40–60	0.01–0.012
Total			7.20–9.82

year^{-1} in finger millet system under diverse management practices. Under *rabi* sorghum production system in Vertisol region of Maharashtra (semi arid) the sequestration rate ranged from 0.1 to 0.29 t ha^{-1} year^{-1} with different integrated management options. In soybean production system in black soils of Madhya Pradesh (semi arid) the potential rate of carbon sequestration is up to 0.33 t ha^{-1} year^{-1} in top 20 cm soil depth (Srinivasarao *et al.*, 2011, 2012a,b,c).

4.9.3 Policy Initiatives for Enhanced Adaptive Capacity

Apart from technological interventions, a sound policy framework and political will are required to battle climate change impacts on agriculture. The policy should address the issues of redesigning social sector with focus on vulnerable areas/populations, introduction of new credit instruments with deferred repayment liabilities during extreme weather events and weather insurance as a major vehicle to transfer risk. Government should identify and prioritize adaptation options in key sectors (storm warning systems, water storage and diversion, drought/pest warning systems and infrastructure needs). State level institutions should be strengthened; a dedicated cell should be set up in each state on close monitoring and take up quick remedial measures in the event of an extreme weather event in agriculture and help in settling insurance claims. Emphasis should be given on tapping financial resources to strengthen adaptation efforts at state level. The capacity of local institutions *e.g.*, SHGs, banks and agricultural credit societies should be strengthened to come up with new way of financing climate related risks at village/block level. Public-private partnership in promoting investments on infrastructure markets, information and communication can help deal with climate risk to a large extent.

5

INTEGRATED FARMING SYSTEMS APPROACH FOR SUSTAINING AGRICULTURAL PRODUCTIVITY

Though crop-based agriculture has helped India to grow at rapid pace during sixties and seventies, the contribution of the allied sectors like livestock, horticulture and fisheries to agricultural GDP is increasing in recent years. The raising labour costs and fluctuating market prices are increasing farmers' risk. Resource degradation problems are also forcing farmers to opt for diversification of enterprises in order to sustain their livelihoods. The Integrated Farming Systems (IFS) approach has the potential to overcome these constraints and achieve sustainable production, higher income and minimize risk. In this section, the most profitable and sustainable IFS models in rainfed and irrigated production system have been discussed based on on-station and on-farm experiences.

5.1 PROFITABLE AND SUSTAINABLE FARMING SYSTEMS IN RAINFED AREAS

Traditionally, farmers in rainfed regions practiced crop-livestock mixed farming system, which provided stability during drought years minimized their risk and helped them to cope with weather aberrations. However, these traditional systems are low productive and cannot ensure livelihood security now. On the other hand, the high productive monoculture systems are also highly risk prone and expose farmers to severe income shocks. Therefore, there is a need to evolve an IFS approach which minimizes risk while ensuring higher returns.

On-station and on-farm research in different rainfed regions of the country has resulted in identification of a number of sustainable and profitable IFS models which can be adopted by the farmers. The main advantages of the farming systems approach involve: (i) *In situ* recycling of organic residues including farm wastes generated at the farm to reduce the dependency on external inputs, (ii) reduction in cost of production through effective use of byproducts and enhanced input use efficiency, (iii) minimizing risk associated with dependence on a single farming enterprise, (iv) conservation of natural resources, and (v) food security, employment generation, enhanced system productivity and income/unit area.

In the IFS approach, several components like crops, livestock, horticulture, poultry and off-farm enterprises get positively integrated at the household level with a view to optimize income and resource use. Farming systems are highly location specific depending on the natural resources available in a given agro-climatic situation. The success of an IFS and its sustainability depends entirely on strengths and weaknesses of the internal factors like natural resources (soil and water, farm size, availability of recyclable residues, labour) and external factors such as market prices, infrastructure and the level of post-harvest processing.

The IFS models in rainfed agriculture can be adopted on a watershed basis or with a farmer centric approach.

5.1.1 Watershed Based Farming Systems

Farming systems approach on a watershed basis is a method designed to understand the farmers' priorities, strategies and allocation of enterprises and decision-making. It should start with analysis of farmers' knowledge, problems and priorities in a given hydrological unit. This approach should relate to the land use that results in an efficient, optimum and sustainable use of natural resources including biotic, socio-economic and related infra-structural resources. The model should address arable, non-arable, common pool resources and private property resources integrating the components of soil and water conservation measures, crops, parkland systems, trees on bunds, wind breaks, silvi-pasture systems, agri-horticulture systems, block plantations, high value low volume crops and live fences in a holistic manner. In this approach, the focus is more on managing natural resources *viz.* soil, water and biodiversity in a topo-sequence, while prioritizing the farming system decisions on an area basis.

In regions with rainfall of 500 to 700 mm, the farming systems should be based on livestock with promotion of low water requiring grasses, trees and bushes to meet fodder, fuel and timber requirements of the farmers. In 700 to 1100 mm rainfall regions, crop, horticulture and livestock based farming systems can be adopted depending on the soil type and the marketability factors. Water harvesting is a major component in this region in the watershed based farming system. In areas where the rainfall is more than 1100 mm, IFS module integrating paddy with fisheries is ideal. There are several modules of rainfed rice cultivation along with fisheries in medium to low lands of rainfed rice growing regions in the eastern states of India.

A model farming system for small holders with 1.15 ha area in an Alfisol watershed has been developed by CRIDA covering arable crops, green manuring, bushes on bunds, fruit plants on the lower side and grasses on the upper topo-sequence in a micro-watershed. Economic analysis of the model after a 3-year period (2005–08) indicated that the module with arable crops on 0.4725 ha, agro-forestry on 0.3496 ha, vegetables on 0.1150 ha, grasses on 0.1256 ha and bushes on 0.0890 ha gave the highest gross income of Rs. 16080/-, net income of Rs. 9793/- and a BC ratio of 2.38. The individual enterprises of arable cropping, agro-forestry, vegetables, grasses and bushes contributed 38.2, 10.3, 27.2, 7.1 and 17.2%, respectively to the total net income (CRIDA, 2008).

Under the technology assessment and refinement programme of the NATP, an IFS module comprising 35.4% area under cereals, 25.7% under pulses, 21% under oilseeds, 17.3% under commercial crops and 1.2% area under fodders along with backyard poultry (6 birds per household) was found to be ideal for small and marginal farmers in Dharwad region of Karnataka (TAR-IVLP, 2005). The poultry component played a major role in stabilizing the farmers' income during drought years. Similarly, in Arjia region of southern Rajasthan, an IFS module of maize, pulses, oilseeds and forage grasses combining with *in-situ* rainwater management and bio-fencing gave 22.37% higher profitability than sole maize (AICRPDA-Arjia, 2006).

5.1.2 Farmer/Family Centric Farming Systems

In this system, the focus is on the individual households particularly of small and marginal farmers and their livelihood security. In this approach, the strengths and weaknesses of the traditional farming systems followed by the farmers need to be assessed based on which selective production enhancement and enterprise diversification interventions are introduced. This calls for a detailed analysis of each household for their resources, investment capacity, and labour availability and so on through participatory farming systems analysis tools. This can be done through focused group discussions, household surveys and local market needs. Based on the income derived from each enterprise by the family (>50%), the family centric farming system can be grouped into crop, agroforestry and livestock based systems (Subba Reddy and Ramakrishna, 2005).

5.1.3 Crop Based Farming Systems

In this system, crops are the main source of livelihood. Animals are raised on agricultural wastes, and animal power is used for agricultural operations and the dung is used as manure and fuel. In case of high rainfall regions, the fish is integrated where the ponds are manured with crop and animal byproducts. Agri-sheep farming involving cotton on one ha of arable land and rearing of 10 lambs gave a net returns of Rs. 27500/ha as compared to growing cotton alone (Rs. 8700/ha) at Warangal district of Andhra Pradesh (TAR-IVLP, 2003). The IFS model in dryland Vertisols at Kovilpatti (Tamil Nadu) showed that Crop + Goat + Poultry + Sheep + Dairy recorded the highest gross income (Rs. 35,301/acre) followed by Crop + Goat + Poultry + Dairy (Rs. 30807), while the conventional system having crop cultivation alone gave only Rs. 5860/acre as gross income. The animal waste from cow (20–22 kg/day/animal), sheep and goat (400–450 g/animal/day) and poultry litter (40 kg/batch of 20 broilers) were collected and applied to the IFS field, which resulted in improved soil fertility and crop yields. Employment increased from 75 man-days in conventional cropping system to 272 man-days in IFS model (AICRPDA-Kovilpatti, 2006). At Chhattisgarh, a model having 2 bullocks + 1 cow + 2 buffaloes + 15 goats + 20 poultry and 20 ducks along with the crops in 1.3 ha gave the net income of Rs. 58,456/year against arable farming alone (Rs. 18,300/year) and employment generation of 571 man-days (Rama Rao *et al.*, 2005). Integration of sheep rearing in groundnut based farming system offered a gainful employment in rainfed areas. In scarce rainfall zone of Andhra Pradesh at Anantapur, the highest net returns were recorded with groundnut cultivation (2 ha) + poultry (Rs. 43,360) followed by groundnut cultivation + dairy (3 buffaloes) (Rs. 40,606) while sole crop of groundnut (2.6 ha) recorded the net returns of Rs. 14872/ha (Reddy, 2005).

In high rainfall areas crop based farming systems will include fisheries. In rainfed rice-based farming system at Orissa, conserving excess water in the refuges at the down stream of rice field and rearing of fish recorded the highest net returns (Rs. 21,197/ha) and BC ratio (2.78) as compared to growing of rice alone (Rs. 15294/ha) (James *et al.*, 2005). In Jharkhand, improved rice (IR-64) + fish (mixed carps) – wheat (PBW-443) enhanced the net returns (Rs. 58557/ha) as compared to the farmers practice of rice-fallow system (Rs. 2770/ha). Also in Jharkhand, Fish-cum-pig rearing (2:3) along with paddy increased the net returns (Rs. 53100/ha) with a B: C ratio of 4.12 as compared to farmers practice of rearing fish alone in the ponds (Rs. 12125/ha) (TAR-IVLP, 2004).

Rice + fish FS in 1.0 ha rainfed lowland area at Central Rice Research Institute (CRRI), Cuttack in coastal Orissa with 13% area for trench and pond refuge and 20% area for dyke all around annually produced 16 to 18 t food crops, 0.6 t fish and prawn, 0.6 t meat, 10,000 eggs, besides flowers, fiber, fuel wood and 5.2 t fodders resulting in net returns of Rs. 55,000 in the first year and about Rs. 1,00,000 in the sixth year, generated 250–300 man days/year and increased productivity and income by 12 to 15 times over traditional rice farming (Sinhababu and Das, 2004).

In another field study on rainfed farming systems in Orissa, integrated farming approach with combination of crops (rice, off-season tomato, cauliflower) and non-crop enterprises like poultry + paddy straw mushroom production + vermicomposting was found superior to farmers' practice (crops alone, rice-green gram) in terms of higher net returns, benefit:cost ratio (1.78) and sustainability index of 54.58% (Table 5.1; Barik *et al.* 2010).

Table 5.1: Economics and employment generation from integrated farming system

Farming system	*Net return (Rs/ha)*	*B:C ratio*	*Sustainability index (%) (Man-days/ha)*	*Employment generation*
Rice-green gram	15820	1.51	9.65	86
crops (rice, off-season tomato, cauliflower) + poultry + mushroom + vermicomposting	44230	1.78	54.58	143
Mean	30025	1.64	–	–
SD	20088	0.19	–	–

5.1.4 Agro-forestry Based Farming Systems

Perennial components like trees and grasses impart stability to farming due to less effect of yearly variation in rainfall on these components besides protecting crops from water and wind erosion and improvement of soil fertility. Several studies carried out at CAZRI, Jodhpur showed higher benefit cost ratio from tree based farming systems as compared to pure arable cropping. Agri-pasture and silvi-pasture systems were found to be more stable and profitable than arable farming. The agri-silviculture system is recommended for land capability class IV with annual rainfall of 750 mm. Short duration dryland crops such as pearlmillet, blackgram and greengram combined with widely spaced tree rows of *Faiderbia albida* and *Hardwickia binata* was found compatible in semi-arid tropical areas (Korwar, 1992). At CRIDA, Hyderabad the horti-pastoral system involving Cenchrus/Stylos in rainfed guava and custard apple, Cenchrus yielded dry forage of 7 t/ha during the first year while stylos recorded 5.6 tons of dry fodder during the second year of plantation. In ber based agri-horti system, pearlmillet + pigeonpea (Solapur), pigeonpea + blackgram (Rewa), castor (Dantiwada) and clusterbean (Hyderabad) showed promising results. At Hyderabad, ber gave about 40 kg fruits/tree along with the 100 kg of horsegram and 450 kg of cowpea cultivated in interspaces (Osman *et al.*, 1989).

5.1.5 Livestock Based Farming Systems

The livestock based farming systems in rainfed agriculture are complex and generally based on traditional socio-economic conditions. An understanding of production factors (livestock, capital, feed, land and labour) and processors (description, diagnosis, technology design, testing and extension) that effect animal production is pre-requisite for livestock integration. The productivity of livestock in farming systems in rainfed agriculture can be improved by increased fodder production as an intercrop with cereals, relay and alley cropping, forage production on bunds, improving the feed value of stover by chopping, soaking with water, urea treatment, strategic supplementation of concentrate, urea molasses etc. Establishment of fodder banks in areas where surplus fodder is available, artificial insemination with semen of approved bulls; removing low grade animals through castration and adoption of preventive measure like vaccination and de-worming through health camps improve the productivity of milch animals (Mishra, 2002). At CRIDA, field studies indicated that urea treated straw increased the milk yield by 0.47–1.2 l/day in IVLP villages of Ranga Reddy district. The paddy straw consumption was also increased with 1–1.2 kg/animal due to this intervention. Urea molasses mineral block (UMMB) enhanced quality and quantity of milk by 25–30% in cows and buffaloes. It also helped in maintaining the overall health and productivity of animal particularly when fodder scarcity was acute during drought period. Mineral supplementation gave higher milk yield (58%) and net returns (Rs. 6816) compared to the farmers' practice of grazing alone (Rs. 2156).

Some of the potential farming systems modules for different agro-climatic zones of the country suitable for rainfed agriculture are given in Table 5.2.

5.2 PROFITABLE AND SUSTAINABLE FARMING SYSTEMS FOR IRRIGATED AREAS

Irrigated areas of the country are facing constraints like decreasing soil fertility, micronutrient deficiencies and accumulation of heavy metals (arsenic, lead, cadmium) in recent years. This is leading to declining productivity and sustainability. Integration of agricultural crops with other enterprises such as horticulture, dairy, fisheries and poultry in a farming system approach has potential to reverse the soil, water and environmental degradation, provide a livelihood

Table 5.2: Rainfed Farming System Modules in different AICRPDA Centers

Production system/ AESR/Soil climate	*Area of domain*	*Potential farming system module*
	Arid Agro-ecosystem	
Groundnut–Anantapur **AESR 3.** Red soils with 669 mm rainfall	*Kharif* arid shallow alfisols with 616 mm annual rainfall, length of growing period (LGP) 60–90 days, Target domain – Rayalaseema (Anantapur, Kurnool and Chittoor districts of Andhra Pradesh)	Groundnut + Pigeonepa (11:1). Crop + livestock (sheep @ 10/ha of farming system) Boundary plantation – *Azardiracta indica, Eucalyptus camaldulensis, Acacia nilotica,* Live fence: Outer layer: *Eucalyptus, Agave;* Inner layer: *Glyricidia*, Fodder production: *D. sissoo, glyricidia,* Fruit trees: ber, custard apple, pomegranate, amla + *kharif* spreading crops, Wood (Commercial/ farm use/fuel wood): *Acacia auriculiformis/Acacia nilotica,* Livestock: Cattle, Buffaloes, Sheep, goat, poultry.
Pearlmillet–Hisar **AESR. 2.3**. Deep aridisols with 320 mm rainfall	Arid deep aridisols covering the areas of Hisar, Sirsa, Fatehbad, Bhiwani, Jhajjar, Mahendergarh, Rewari, Gurgaon, Kandi area of Panchkula and Ambala	Chickpea paired row (30/60 cm) + Chinese cabbage, ber, jatropha + aswagandha/Aloe vera Boundary plantation – *Prosopis cineraria, Acacia tortolis, Eucalyptus* Live fence–Outer layer: *Prosopis juliflora, Zizyphus mauritiana, Zyziphuz numularia,* Inner layer: *Dalbergia sissoo, Carissa carundus*
	Semiarid Agro-ecosystem	
Rabi sorghum–Bijapur **AESR. 3.** Semi-arid medium/deep vertisols with 680 mm rainfall	Semi-arid medium/deep vertisols covering domain districts of Bijapur, Bagalkot and Gulbarga, eastern parts of Belgaum, Lingsugur of Raichur of Karnataka and Southern parts of Maharshatra	Agri-horti system of *rabi* sorghum + ber, subabul *Acacia nilotica* and *Dalberja sissoo* on degraded soils Pigeonpea + aswagandha/ sunflower + aswagandha, Amla/custard apple + henna + castor/ pearlmillet, Sapota + pearlmillet/castor, Guava + henna + Seemarisba + grengram Boundary Boundary Plantation – *Eucalyptus camaldulensis, Tectona grandis* Live fence – Outer layer: *Agave sisalana,* Cactus; Inner layer: *Leucaena leucocephala, Carrissa carundus,* Trees on crop lands–Fodder/green biomass: *D. sissoo,* Glyricidia, *Albizzia lebbeck, Hardwickia binata* Fruit: Mango, Pomegr-anate Wood (Commercial/farm use/fuel wood): *Eucalyptus hybrid, Acacia auriculiformis*

Table 5.2: (*Contd..*)

Table 5.2: (*Contd..*)

Production system/ AESR/Soil climate	*Area of domain*	*Potential farming system module*
***Kharif* sorghum–Hyderabad** *Kharif* rainfed alfisols with 750 mm annual rainfall, LGP 120–150 days	Southern Telangana zone (Ranga Reddy district, Nalgonda, Mahaboobnagar, Warangal and parts of Medak district).	IFS model for 2 ha: sorghum + pigeon pea (0.4 ha), castor + cluster bean (0.4 ha), sunflower + pigeon pea (0.3 ha), *Cenchrus* + *Stylo* (0.5 ha), Amla/mango (0.2 ha) and vegetables (0.2 ha) integrating with backyard poultry/apiary/small ruminants, together with water harvesting pond
	Sub Humid Agro-ecosystem	
Rice - Phulbani **AESR 12.1.** Red and yellow soils with 1597 mm annual rainfall	Sub-humid Deep Alfisols/Oxisols covering areas of Kandhmal, Boudh, Rayagada, Gajapati and part of Ganjam districts *i.e.* Phulbani, Boudh, Rayagada, Gajapathi and Asoka	Rice + Pigeonpea (5:2) Maize + Pigeonpea (1:1) Mango + blackgram/niger agri-horti system Pigeonpea + turmeric yam (Diascorea) + maize Sal, Teak in wastelands, Boundary Plantation – *Gmelina arborea, Tectona grandis* Live fence – Outer layer: Agave Inner layer: *Glyricidia*, Sesbania Trees on crop lands–Fodder/green biomass: *Cassia* sp, Fruit: Mango, Guava, Lime Tuber crops: Colacassia, Cassava,Yam Wood (Commercial/ farm use/fuel wood): *Eucalyptus* Spices: Turmeric, Ginger
Rice-Ranchi **AESR. 12.3.** Red light textured soils with 1500 mm rainfall	Sub-humid Alfisols covering entire plateau of Jharkhand state, parts of Rohtas, Gay, Jamui in Monghyr district, Banka sub-division of Bhagalpur district in Bhir and Purulia and Bankura districts of West Bengal	Rice+ pigeonpea (3:1) and groundnut + pigeonpea (2:1) in uplands Agri-horticulture: Fruit crops (Mango /citrus/sapota / pomegranate/custard apple/litchi/jack fruit/jamun) + field crops (pulses/oilseeds) Alley cropping: *Leucaena* + turmeric/ginger Boundary Plantation – Eucalyptus Live fence–Outer layer: *Lawsonia* Inner layer: *Leuceana, Tectonia grandis,* Trees on crop lands–Fodder/green biomass: *Leucaena* Fruit: Mango, Ber Wood (Commercial/farm use/fuel wood): *Tectona grandis, Eucalyptus* Medicinal and aromatic plants: Vetivera, Palmrosa Livestock: Cattle, Buffaloes, Sheep, Poultry, Duckery, Piggery.

diversification and continuous flow of income and employment throughout the year.

5.2.1 Agro-forestry Based Farming Systems

Integration of multi-purpose trees with crops has good potential for value addition, income and employment generation. Promising agro-forestry trees for irrigated areas include: poplar, eucalyptus, teak and *drake*. The poplar-based agro-forestry was relatively more remunerative and gave higher returns/annum compared to arable cropping in the Indo-Gangetic plains (Dhanda, 1999). A five year old poplar plantation in riverine land of Punjab can yield about 240 t/ha timber wood of worth Rs. 4,80,200. In addition to economic benefits, the tree-based systems also improve soil fertility through build-up of organic matter (Singh *et al.*, 1997; Kaur *et al.*, 2001) and nutrients in the soil. In the black and red soil region, the land scape is often undulating. The canals run on the ridge. The seepage water travels through the porous *murrum* layer to areas at lower elevation, picking enough salts on the way to salinize lands on the slopes or in the valley. A belt of agro-forestry plantations involving trees capable of transpiring large amounts of water can effectively intercept such saline seeps and protect highly productive valley lands. *Acacia nilotica* and *Dalbergia sissoo* account for about 86 and 84% interception of the seepage over the control, respectively (Patil *et al.*, 1994).

However, the popularity of tree-based systems is largely governed by the market infrastructure and price trends. *Eucalyptus tereticornis* based agro-forestry became quite popular in early eighties and large scale plantations came up as boundary plantations and block plantations. By the time these plantations became ready for harvest in early nineties, the prices crashed to all time low. Farmers harvested Eucalyptus and sold at through away prices. This markedly affected farmers' interest in agro-forestry. The same scenario happened with *Kinnow* and grapes in Punjab. The recent price trends of the most dominant poplar based agro-forestry in Punjab, Haryana and western UP is a reflection of what happened with Eucalyptus in early nineties. To promote adoption of such diversified agro-forestry systems, policy initiatives like assured procurement, post-harvest and value addition are needed.

5.2.2 Multi-enterprise Farming Systems

***Crop-Poultry-Fish system*:** In the lowland rice growing areas of Tamil Nadu, two rice crops (June-September and October - January) followed by a rice fallow or green manure (February - May) is a common practice. A new system was formulated to link poultry-cum-fish culture as multi--enterprise option (Shanmugasundaram *et al.*, 1995). The crops raised under this system were rice (0.96 ha) - rice (0.96 ha)--cotton (0.76 ha) + maize (0.20 ha). Maize grain formed major source of poultry feed. An area of 0.04 ha was allotted for fish pond over which poultry shed was constructed in one corner. Crop based system gave an annual net income of Rs. 8312/ha, modified cropping system Rs. 15,009/ and crop + poultry + fish system gave Rs. 17,209/ha.

Integrated Farming System involving Rice-Fish-Duck

Similarly in Karnataka, integration of rice + fish with poultry yielded the highest production of 11.13 t/ha. It also produced 42% higher energy output than rice-rice system. In another study in Tamil Nadu, the highest net returns of Rs. 1,31,118/ha were obtained by integrating goat + fish + cropping applied with re-cycled fish pond silt fed with goat droppings (Jayanthi *et al.*, 2001). In comparison to 369 man-days/ year generated in cropping alone system, the IFS system generated 576 man-days/year (Table 5.3).

Table 5.3: Productivity and economic analysis of integrated farming systems in Tamil Nadu

Farming systems	*System productivity (Rice eq. yield t/ha)*	*Net returns (Rs/ha)*	*B:C ratio*	*Per day returns (Rs.)*	*Employment generation (man-days /year)*
Cropping alone	13.0	37,153	2.43	178	369
Cropping + fish + poultry	29.6	97,731	3.02	400	515
Cropping + fish + pigeon	29.2	98,778	3.06	400	515
Cropping + fish + goat	37.7	1,31,118	3.36	511	576

***Crop-duck-fish system*:** A series of experiments conducted in south India particularly at coimbatore clearly indicated the superiority of integrating aquaculture and duck with cropping (Table 5.4). The droppings of the ducks are effectively recycled for manuring the fish pond (Ganesan *et al.*, 1990).

Table 5.4: Economics of rice-based duck-cum-fish culture in Tamil Nadu

Details	*Gross returns (Rs/ha)*	*Net returns (Rs/ha)*	*Employment generation (man-days)*
Conventional cropping system (Rice-rice-rice-fallow-pulse)	25625	13790	252
Mixed farming system (Rice-rice-rice-fallow-cotton+maize and duck-cum-fish)	43185	24117	396
Additional income/employment over conventional cropping system	17560	10327	144

***Rice-fish culture for deep water rice growing areas*:** This multi-enterprise option is generally followed in coastal wet lands particularly deep water rice growing areas. Among the rice-based multi-enterprise options tried, rice-rice-azolla and rice-azolla-calotropis + fish farming gave higher grain yield of 10.57 t/ha and 10.13 t/ha^{-1}, respectively. On an average, 173 kg/ha fish was harvested from rice-rice-azolla-calotropis+fish farming (Shanmugasundaram and Balusamy, 1993).

***Crop-dairy system*:** Integration of dairy (one or two milch buffalows on the farm) with sugarcane-wheat system is quite remunerative as this rotation generates sufficient good quality fodder as byproducts (PDCSR, 2002). The comparative advantage of integrating dairy with cropping in Meerut area of Uttar Pradesh on different categories of farm size is given in Table 5.5.

Table 5.5: Economics of dairy-based farming system in Meerut, Uttar Pradesh

Farm size	*Net returns (Rs/ha)*	
	Sugarcane-wheat	*Sugarcane-wheat + dairy*
Marginal	38750	45950
Small	39105	45825
Medium	43785	51225
Large	42430	47950

There are large areas of waterlogged fertile alluviums in eastern India (3.28 M ha) where water stagnates above ground for over six months in a year. The adverse physical conditions allow only one anaerobic paddy crop with a very low yield potential of less than 1.0 t/ha. A number of farming models integrating fish and prawn culture, cultivation of paddy, vegetables, fruits, poultry, duckery, piggery, rabbitry and cattle were evaluated at different locations. The systems based on multiple recycling of carbon, energy and nutrients from biomass to livestock/poultry/piggery/fishery etc also minimized environmental loading with pollutants (Samra *et al.*, 2003). The average net return of Rs. 69,275/ha/year from integrated farming system were higher than traditional rice cultivation. Such enterprising ventures need to be promoted in Orissa, Bihar, Assam, West Bengal and Eastern Uttar Pradesh for enhancing livelihood and land quality.

5.3 PROFITABLE AND SUSTAINABLE FARMING SYSTEMS FOR HILL ECOSYSTEM

Hill farming is a self contained type of system in which land, livestock and people are closely linked (Fig. 5.1). The system is predominantly sustained by a flow of organic material and plant nutrients from uncultivated land and adjoining forests. Supply of this crucial source is getting jeopardized due to over felling and over grazing with the result crop and fodder production per unit time have decreased over the years.

However, the fragile eroded hill soils are extremely low in available nutrients and organic matter content, incapable of sustaining increasing demand for food and fodder caused by increased population of both human and animals. As a result, at present an average farmers' family produces food grains just sufficient to meet 4–5 months of its requirements. The hilly regions of the country are today confronted with multiple of soil and water resources related problems like increased sedimentation of rivers and water bodies, recurring floods, drying up of water sources and above all the deteriorating socio-economic status of the people. Therefore, the above depicted model is very common for subsistence agriculture.

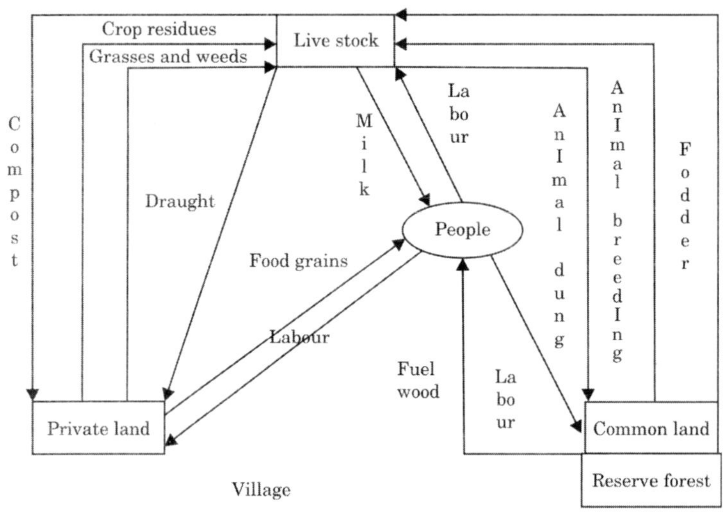

Fig. 5.1: A conceptual model showing close link between land-livestock and people in hilly area (*Source*: Gill, 2010)

Prasad (1993) gave guidelines to identify agricultural, horticultural, livestock and agri-horti-silvi pastural based farming system research approach for the NEH hill region of India. Growing crops with fruit trees is highly economical and liable to minimize the risk factor occurring with sole cropping under low rainfall. Trees help in nutrient pumping from lower soil strata to the crop root zone.

Baishya *et al.* (2007) reported that crop + dairy + fishery + poultry, crop + dairy + goatery and crop + dairy + goatery + pigeon were the remunerative systems for marginal, small and medium farmers under rainfed situations of Kamrup district in Assam.

The hilly terrain of NE hill region is suitable for sustainable multi-enterprise system. The Tripura center of this region has developed a farming system model combining agriculture with horticulture, forestry and livestock rearing on one hectare land. The enterprises taken are: cereal crops, pulses, oilseeds, horticultural crops such as mango and pineapple, vegetables, and livestock components of duckery, piggery and fishery in the water harvesting structures. The results indicated that the multi-enterprise system is nearly five times more profitable than traditional monocrop rainfed rice cultivation.

5.4 PROFITABLE AND SUSTAINABLE FARMING SYSTEMS FOR COASTAL AREAS IN EASTERN INDIA

The coastal areas of currently low productivity offer tremendous opportunities to be harnessed by adopting proper strategies by well focused action plans under IFS approach such that agriculture will be more productive, profitable, sustainable, competitive and eco-friendly. In coastal India, land holdings are small and fragmented, operational resources are scarce and rice continues to be the major crop. Nanda and Garnayak (2010) have reviewed the work done on integrated farming systems for livelihood security in coastal ecosystems. Coastal ecosystem represents a transition from terrestrial to marine influences and vice versa. It comprises not only the shoreline ecosystem but also the upland watersheds draining into the coastal waters and the near shore sub-literal eco-systems influenced by the land-based activities. The production constraints in coastal areas are: soils are deep to very deep, poorly drained, sandy to fine loamy and subjected to the influence of tidal waves and periodic inundations, shallow water table enrich with salts, heavy rainfall and cyclonic weather, agriculturally poor lands and mono-cropping of rice in waterlogged areas (Nanda and Garnayak, 2010).

Rice based FS is the single most important source of income and employment for the majority of the people in this tract and is a viable option for meeting their livelihood support. Rice based FS such as rice + fish, rice + duck, rice + fish + duck, rice + fish + *Azolla,* rice + fish + poultry/duck + vegetables + fruits + agroforestry are in practice in the coastal areas. In rice ecosystem, the rice bottom is highly fertilized on account of the production of zooplankton and phytoplankton and these resources are fully utilized by fish. Free movement of fish in water body loosens and aerates the soil, enhances the decomposition of organic matter and promotes release of nutrients from soil. The unutilized fish feed, a decayed *Azolla* and fish excreta settled at the fish trench bottom has a higher nutritive value (Balusamy *et al.*, 2003). Rice-fish-duck culture has great potential in coastal ecosystem and need to be encouraged particularly in the rainfed lowland areas. Ducks are eco-friendly who used to destroy harmful weeds, snails and insects from rice fields and improve soil fertility. Duck droppings serve as feed for fishes. Thus fish and duck together would ensure high profit due to higher fish yield, duck eggs (280–300/year), duck meat and enhanced rice yield (8–10%).

The coastal regions receive high rainfall and excess rainfall can be stored in ponds that would help in raising a second crop after rice and also fish. That is why pond based IFS encompassing fish-rice-cow/pig-duck/poultry-vegetables are feasible particularly, in rainfed lowlands. In this system, marginal land/wet lands are brought into productive use, animal wastes are used to fertilize fish pond/refuge and crop lands, while the lands in return produce crops which serve as food for animal, fish and men and the pond provides food for fish, duck and men. As system's dykes are comparatively wide, housing units for cow/pig/duck/poultry along with cultivation of economic varieties of plants such as vegetables and fruit plants are carried out, using pond water for life saving irrigation at different growth stages. Profit of margin will be enhanced much more than any individual system. Besides, this can provide gainful employment to family members and improve their nutritional status. The dykes to be constructed for preventing escape of fish from the integrated system can be used for growing vegetables and other fruit trees like papaya and banana to make the system more economically viable.

Vegetables such as gourd, radish, brinjal, okra, leafy vegetables during *kharif* season and vegetables such as tomato, french beans, radish, bitter gourd, cucumber, cauliflower, cabbage, brinjal, pumpkin and leafy vegetables (coriander, amaranthus and Indian spinach) can be grown during winter. Vegetables such as snake gourd, bitter gourd, ridge gourd, bottle gourd or ash gourd can be grown throughout the year on raised platforms and *okra, poi,* cowpea, bitter gourd, ridge gourd and pumpkin in the field during summer season.

Rice + fish FS in 1.0 ha rainfed lowland area at Central Rice Research Institute (CRRI), Cuttack in coastal Orissa with 13% area for trench and pond refuge and 20% area for dyke all around annually produced 16 to 18 t food crops, 0.6 t fish and prawn, 0.6 t meat, 10,000 eggs, besides flowers, fiber, fuel wood and 5.2 t fodders and resulted in net returns of Rs. 55,000 in the first year and about Rs. 1,00,000 in the sixth year, generated 250–300 man days/year and increased productivity and income by 12 to 15 times over traditional rice farming (Sinhababu and Das, 2004).

Similarly, intensive rice-fish farming in 0.4 ha area under irrigated condition with 60% area for growing rice (3 crop sequence) + fish and 40% bund area for vegetables, pulses, fruits (150 banana, 60–70 papaya, pineapple, guava, mango), trees (teak, sisoo, neem, aonla) spices and condiments, mushroom (4–6 units), flower and duckery (20 ducks), dairy (single cow) produced 12 to 18 t food crops, 0.3 to 0.4 t fish, 3 to 4 t rice straw thereby increasing productivity by 15 to 18 times over conventional farming. The net returns in the first year was Rs. 15,000 which doubled in third year besides the expected income of Rs. 5,00,000 from timber and other trees in the long run. The system also ensures sustainable household food security with monthly income of Rs. 3,500–4,000 and employment generation of 450–500 man days/year (Jha *et al.*, 2001). A multi-tier rice-fish-horticulture-agroforestry based FS in 0.8 ha of deepwater area at the same location integrating components like, wet season rice followed by other season crops and rice in dry season, fruit crops, tuber crops, vegetables, agroforestry, poultry and duckery produced about 8 t/ha of rice (two crops), 2–27 t/ha of vegetables in uplands (tier II), 10–25 t/ha in tuber crops in uplands (14.0 t/ha of sweet potato under zero tillage in lowland rice fallow), 10–70 kg/plant of fruit crops in upland (tier I), 1.0 t/ha of fish and prawn and 0.5 t of meat. The net income was around Rs. 1,00,000/ha/year.

IFS in 2.47 ha water logged area in coastal tract of Orissa with 1.64 ha converted to grow-out fish pond and rest 0.83 ha as raised embankment all around the pond for growing vegetables, fruits, flowers and the poultry shed resulted in productively of 9.42 t fish equivalent/ha with gross and net returns of Rs. 6,51,110 and Rs. 3,62,515, respectively during 2002 as compared to an average yield of 3.5 t rice/ha from the conventional low land rice (Samra *et al.*, 2003).

Net returns from different fish-based FS such as fish-poultry, fish-duck and fish-pig each in 1.0 ha pond at Central Institute of Freshwater Aquaculture (CIFA), Bhubaneswar were Rs. 1,37,200, 1,16,126 and 1,72,540 with B:C of 1.42, 1.62 and 1.44, respectively. The systems include stocking density of 5,000–6,000 carp fingerlings, 500–600 poultry, 200–300 ducks and 30–40 pigs (CIFA, 2009). A pond based IFS in 1.25 ha area at Bhubaneswar comprising of field and horticultural crops, fishery, poultry, duckery, apiary, mushroom, dairy and agroforestry generated net income of Rs. 58,360, employment of 573 man days with B:C of 2.18 ensuring high standard of living for small and marginal farmers (Behera and Mahapatra, 1999). Later, Behera *et al.* (2008) reported net returns of Rs. 47,825 with employment generation of 248 man days from an on-farm IFS (0.4 ha) comprising of field and horticultural crops, fishery, dairy and agroforestry under North Eastern coastal plain zone of Orissa, which were 87 and 82% higher respectively than the conventional practice involving crop component only.

5.5 PROFITABLE AND SUSTAINABLE FARMING SYSTEMS FOR COASTAL AREAS IN WESTERN INDIA

The work done on integrated farming systems for livelihood security in coastal ecosystems of Western India has been reviewed by Korikanthimath and Manjunath (2009); Varughese and Mathew (2009) and Singh and Manjunath (2010). ICAR Research Complex for Goa has identified different farming system models involving rice, coconut, etc. involving the possible crop combinations and integration with livestock and fish species. Research efforts of this institute have clearly indicated that intercrops in coconut with high yielding forage grass–legume mixture and integration with dairy through recycling of manurial resources are productive and profitable. Further, the generation of employment was also substantial. Package of practices for similar systems involving rice- mushroom and poultry, rice-fish, duck-fish, poultry-fish systems are also developed at the ICAR Research Complex for Goa. The results clearly show the supremacy of system approach with increased productivity, employment generation and profitability.

Adoption of an intensive cropping system like rice-brinjal could bring in additional employment opportunities to the tune of 118 man-days/ha over mono-cropping of rice (Table 5.6; Korikanthimath and

Manjunath, 2009). Thus, the family labour get an additional employment on their own farm, that too in the off season which could earn exactly double the wages (Rs. 15,440/ha) for labour employment compared to growing of rice alone (Rs. 7,720/ha). Further, when these intensive cropping systems are integrated with labour intensive enterprises like poultry and mushroom, an additional 164 mandays of employment could be expected. Thus, the total labour employment opportunities were enhanced three and half times within the same farm. Rice-cowpea + mushroom + poultry and rice-groundnut + mushroom + poultry farming systems are also equally good in enhancing the manpower absorption capabilities on the farm which recorded 213.6 and 218.2% more man-days over control. The additional labour employment opportunities of the farming systems could be attributed to labour intensive nature of the integrated enterprises like poultry which alone contributed 24.2% of the man-days.

Table 5.6: Economics and employment generation of rice based farming systems

Farming system	*Net return (× 10^3 Rs/ha)*	*Employment generation (Man-days/ha)*
Rice cropping alone	19.21	110
Rice-groundnut + mushroom + poultry	60.65	350
Rice-cowpea + mushroom + poultry	73.43	345
Rice-brinjal + mushroom + poultry	77.31	392
Rice-sunhemp + mushroom + poultry	52.75	309

Rice + fish FS has great relevance to the coastal low land rice areas of Kerala where the entire food chain and vast amount of fertile water can be fully utilized by integrating rice with fish. The cost of cultivation of rice in this system was reduced by 29% and the cost of production by 34% as compared to mono cropping of rice. The major returns (79%) from a rice-based farming system in Western part of Kuttanada ecosystem of coastal Kerala were obtained from coconut-banana inter-cropping in the dykes and field bunds (Varughese and Mathew, 2009).

Coconut-Based Farming System (CBFS) involving cultivation of compatible crops in the inter-space of coconut and integration with other enterprises like dairy, apiary etc offer considerable scope for increasing productivity and employment. The CBFS is an important farming practice among the small and marginal farmers to maximize land use, farm productivity and profitability. A number of researchers have confirmed the suitability of CBFS in coconut fields (Nair, 1979; Bavappa, 1990).

The crown of the coconut tree, particularly its morphology and palm height influence the available PAR for the under storey crops (cereals, root, fruits, pulses and fodder crops) which are identified for intercropping in coconut gardens and coconut based homesteads (KAU, 2007). Thus CBFS provides food security through food sufficiency and nutritional foods. Intercropped cereal as a source of carbohydrates, proteins and fats, root crops as source of carbohydrates and minerals, pulse crop as protein, fat and oils, fruit crop as vitamins and mineral, vegetable crop as vitamins, minerals, spice crops as food flavoring, vitamins and mineral and cocao as beverage are essential for food security.

The coconut-based IFS studies conducted at CPCRI, Kasaragod and Kayamkulam revealed that the grass-legume fodder from coconut garden in the ratio of 3:1 (30 to 40 kg) could support 4 milch cows/ha. The total output from 1.04 ha IFS model in 1994–95 was 19,125 nuts, 9,275 liter milk, 526 kg poultry, 50 Japanese quail bird and 400 kg fish. The economic analysis revealed that the total variable cost for the system was Rs. 1,59,939. The gross and net returns/annum from the system was Rs. 2,07,169 and Rs. 66,678 respectively (Reddy and Bidappa, 2000). The IFS with small farmers (2 ha and less) developed through linear programming comprising 43 enterprises and with a cropping intensity of 161 provides a cost benefit ratio of 1:2.5 (Jacob and Nair, 1998). The crop livestock components selected in the models interact synergistically to increase the productivity and generate higher net returns.

Similarly, in another experiment on CBFS at Goa, in contrast to monocropping of coconut (77 mandays), the integrated system [coconut + grass + dairy (2 milch animals)] involved an additional 279 mandays of employment, which when engaged by the family labour, could lead to an additional income of Rs. 23,780/ha. Further, integrated dairy provided intermittent returns and regular employment throughout the year for maintenance of cows and forage production (Table 5.7; Korikanthimath and Manjunath, 2009).

Table 5.7: Economics and employment generation of rice based farming systems

Farming system	*Net return (× 10^3 Rs/ha)*	*Employment generation (Man-days/ha)*
Coconut alone	6.23	77
Grass intercrop alone	19.35	219
Coconut + grass intercrop	25.58	137
Coconut + grass + dairy	32.34	356

Arecanut as a sole crop does not utilize fully the natural resources *i.e.* soil, space and light. Arecanut plantations are more ideal for inter/ mixed cropping and multi storeyed cropping than coconut plantations. The compact nature of arecanut crown raised well above the ground (10 to 15 m), allows more sunlight to pass down to ground and maintain high humidity which in turn favours excellent growth of shade loving shrubs/trees (banana and cocoa) and vines (betel leaf and pepper) and variety of short statured biennials and annuals (pineapple, ginger, turmeric, yams and other tuber crops). Cardamom has been one of the time tested most remunerative spice crop in arecanut gardens. Returns as high as Rs. 1,61,840/ha could be obtained by mixed cropping of arecanut and cardamom as against Rs. 1,03,625/ ha (mean of 8 years) in case of sole arecanut. In recent years, Vanilla as mixed crop with arecanut has become the most profitable spice (Korikanthimath and Manjunath, 2009).

Similarly, it is possible to grow such intercrops as pineapple, papaya, annual drumstick, cucurbitaceous vegetables, groundnut, cowpea etc in cashew garden. This is the most important practice to tide over the gestation period of the garden. Intercrops like cereals, groundnut, cowpea, vegetables like chilies, brinjal, amaranthus, flower crops like marigold etc. can be taken up with protective irrigation in the *Rabi* season. Cashew trees can also be planted all along the bunds of upland rice fields (Desai and Adsule, 2002). Possible cashew based inter/mixed crops are: cashew + fruit crops, cashew + pineapple, cashew + guava, cashew + spices, cashew + kokum (*Garcinia indica*), cashew + turmeric, cashew + vegetable crops, and cashew + drumstick. There is great scope for integration of enterprises like goatery, apiary, rabbitry, back yard poultry, piggery, etc., along with cashew depending on the resources in the region Singh and Manjunath (2010).

In a trial on duck-fish integrated farming system for Goa, about 70 to 100 m^2 area was needed as night shelter to accommodate 200 to 250 birds which would cater the need of one hectare pond. With 200 to 250 birds about 10–12 t droppings were recycled/year which was sufficient to meet the manure requirement for one hectare pond. Further, for poultry-fish integrated system in Goa, four varieties of fish namely catla, rohu, mrigal and common carp were stocked in the ratio of 2:2:1:1 at the stocking density of 6000/ha. Poultry shelter of the size of 100–200 m^2 area was required to accommodate 400–500 birds which were sufficient to fertilize a pond of one ha size. A fish production rate of 1,250 kg/ha/8 months was achieved with an average individual growth of one kg when stocked with advanced carp fingerlings at 1500–2000/ha stocking density and without supplementary feeding and pond fertilization (Anon, 1997).

5.6 ADDING FURTHER VALUE TO FARMING SYSTEMS APPROACH

Programmes promoting farming systems should have a built in component of improving soil organic matter. Composting methods with high biomass-to-dung ratio should be targeted to overcome the limitation of availability of dung. A regular subsidised transport (preferably through bullock carts) for manures to the distant agricultural fields should be provided. There is scope for integrating this service with MNREGS. Further, provision of a power operated biomass-shredder as a common utility at the village level would help in cutting the biomass for faster decomposition in manure pits. Such a facility would also increase the fodder supply many fold by reducing wastage and chaffing the hard stumps.

Water sources are primarily groundwater in areas with less than 750 mm rainfall. Efforts should be made to provide incentives to farmers owning groundwater sources to share water with other rainfed farmers to provide protective irrigation during *Kharif* season. Similarly, Pooling/leasing of bore wells or taking over rights on bore wells (at least for *Kharif* season) may be possible if packaged with right incentives. Subsidised energy or renewable energy systems may also be used as an incentive for water sharing and social regulation.

Isolated patches of forest land under ownership of the forest department within the villages/watersheds needs special dispensation. Owing to smaller size, high human and livestock pressure, they are highly degraded and are not brought under any management regime. Regenerating these lands with people's participation for providing biomass for livestock and livelihoods should be the core purpose of managing these forest areas. The community land in the villages, which is accessible for better use, must be used for productive purpose. Therefore, adoption of concept like social forestry, water harvesting and recycling, fishery and stallbed feeding to the animals (goatry/ piggery) will add to the profit margin with other numerous indirect benefits of employment and to improve the ecology of the area.

Biomass intensification specially targeting the small ruminants should receive highest priority; much of the shrub/tree-biomass for goats and sheep can be enhanced easily with little effort/resources. Similarly, adequate investments should be made on community-managed livestock health care systems with strong linkages with animal husbandry departments.

Post-harvest processing and establishing market linkages add further value to the farming systems approach. Collective procurement and marketing if directly handled by the producers will significantly enhance the profitability. With the growth in SHG movement across the country, options for collective marketing and value addition are opening up. This unique opportunity should be harnessed by dovetailing required processing infrastructure and technologies. A special area based planning exercise for mapping the requirements of processing/value adding infrastructure need to be taken up. The infrastructure should include common storage places for seeds and other agriculture inputs and agriculture produce within the village and at the bulking points.

Establishment of agro-service centers in the villages can save the cost of inputs and can also get timely farm advisory services for higher profitability. An effective rural knowledge society and ICT system involving various stakeholders – farmers, development agents and agencies, knowledge generators and distributors (universities and public and private institutions) should be established for bridging the information and knowledge gaps on production and marketing of different commodities.

Formation of commodity groups and self-help groups of farm women can help to promote off-season income generation activities which lead to livelihood improvement. Imparting training in emerging skills and crafts as per diversified demands in the market will go a long way for realizing all inclusiveness of landless, asset-less, small and marginal farmers.

Community level seed banks with buffer stocks of seed material of diverse crops appropriate for the village/area need to be maintained. These seed banks should be considered as a necessary common infrastructure for rainfed farms supported by the government on a regular basis. Seed banks should be controlled and maintained by organized farmer groups. Proper tie-up of the seed banks with seed farmers needs to be established.

The approach of family focused development through tree based farming integrated with other on-farm and off-farm interventions is popularly known as "Wadi" (Gujrati term for orchard). Presently, the wadi program is benefitting over 100,000 families spread over Gujarat, Maharashtra, Karnataka, Rajasthan, Uttar Pradesh and Madhya Pradesh. The model evolved over a two decades period became a scalable approach for sustainable livelihood and better quality of life.

6

CONCLUSIONS

Prudent management of natural resources is key to sustaining agricultural productivity. Technologies for efficient use of natural resources like soils and water are complex and require dedicated effort for adoption at farm level. Many of these technologies are to be applied at watershed or land scape level and therefore, community oriented extension systems have to emerge in India. Managing rain water and ground water are key for supporting future food grain production in India. With competing uses for land and water, a suitable policy instrument is urgently needed for protecting fertile agricultural lands and water resources being diverted away from agriculture. Climate change and climate variability are expected to impact natural resources quality which puts further pressure on governments and farmers to invest more on resource management. To ensure stability of production and farm income integrated farming systems approach is essential both in irrigated and rainfed regions.

REFERENCES

Abrol, I.P. and Sangar, Sunita 2006. Sustaining Indian Agriculture – conservation agriculture the way forward. *Current Science*, 91(8): 1020–1025.

Aggarwal, P.K. 2009. Global Climate Change and Indian Agriculture: Case Studies from the ICAR Network Project. ICAR Publication, p. 148.

Aggarwal, R.K. and Kumar, Praveen 1993. Sustainable productivity through organic manures in arid land ecosystem. Proc. of International Symposium on Environmental Degradation, CAZRI, Jodhpur.

Ahmad, N., Qureshi, R.H. and Qadir, M. 1990. Amelioration of a calcareous saline-sodic soil by gypsum and forage plants. *Land Degradation and Rehabilitation* 2: 277–284.

AICRPDA-Arjia 2006. Annual Progress Report of All India Coordinated research Project for Dryland Agriculture, Arjia, MPUA & T, Udaipur.

AICRPDA-Kovilpatti, 2006. Annual Progress Report of All India Coordinated research Project for Dryland Agriculture, Kovilpatti, TNAU, Tamil Nadu. pp. 31–38.

AICRPDA, 2003. Annual Reports of 22 centers 1971–2001. Eldoscope. Electronic document. All India Co-ordinated Research Project for Dryland Agriculture (AICRPDA), Central Research Institute for Dryland Agriculture (CRIDA), Hyderabad, p. 76357.

All India Coordinated Research Project on Dryland Agriculture (AICRPDA) 1999–2003. Annual reports for the year 1999 to 2003. Central Research Institute for Dryland Agriculture, Hyderabad.

Anonymous, 1976. National Commission on Agriculture, Ministry of Agriculture and Irrigation, Government of India, New Delhi.

Anonymous 1997. *Vision-2020*. Perspective plan of ICAR Research Complex for Goa, Old Goa, p. 66.

Anonymous, 2008. Harmonization of wastelands/degraded land Datasets of India. Published by National Rainfed Area Authority (NRAA), Ministry of Agriculture, Government of India, NASC complex, DP Shastri Marg, New Delhi - 110 012, p. 5.

Annual Report. All India Coordinated Research Project on Water Management for 1972–73 to 1996–97.

Annual Report. All India Coordinated Research Project on Water Management, 2009–10.

Annual Report. All India Coordinated Research Project on Groundwater Utilization, 2009–10.

Baishya, A., Kalita, M.C., Mazumdar, D.K., Hazarika, J.P. and Ahmed, S. 2007. Characterization of farming systems in the Borpeta and Kamrup districts of lower Brahmaputra valley zone of Assam. *Journal of Farming Systems Research and Development* 13(2): 168–175.

Bajwa, M.S., Singh, B. and Singh, P. 1993. Nitrate pollution of ground water under different systems of land management in Punjab. pp. 223–230 in Proceedings of First Agriculture Science Congress, 1992, New Delhi, India: National Academy of Agricultural Sciences.

Balusamy, M., Shanmugham, P.M. and Baskaran, R. 2003. Mixed farming an ideal farming. *Intensive Agriculture* 41(11–12): 20–25.

Barik, K.C., Mohanty, D., Nath, S.K., Sahoo, S.K. and Soren, L. 2010. Assessment of integrated farming system for rainfed ecosystem in Deogarh district of Orissa. *In*: Extended Summaries, 2010. XIX National Symposium on *"Resource Management Approaches Towards Livelihood Security"*, held during December 2–4, 2010 at Bangalore. p. 172.

Bavappa, K.V.A. 1990. Agronomic desirability of coconut based farming systems. *In*: Coconut-based Farming Systems. S.de.Silva (*Ed.*) Jakarata. Asian and Pacific Community. pp. 57–68.

Biradar, P.P., Aladakatti, Y.R., Rao, T.N. and Tiwari, K.N. 2006. Site-specific nutrient management for maximization of crop yields in Northern Karnataka. *Better Crops*, 90(3): 33–35.

Bos, M.G. 1980. Irrigation efficiency at crop Production Level, ICID Bulletin 29(2): 18–26. New Delhi, India.

Chaudhury, J., Mandal, U.K., Sharma, K.L., Ghosh, H., and Biswapati Mandal. 2005. Assessing soil quality under long-term rice based cropping system. *Communications in Soil Science and Plant Analysis* 36: 1141–1161.

Chaudhury, P.R., Rai, P., Patnaik, U.S. and Sitaram, R. 2005. Live fencing practices in the tribal dominated Eastern Ghats of India. *Agroforestry Systems* 63(2): 111–123.

Christiansen, J.E. 1942. Hydraulics of sprinkling systems for irrigation. *Transactions of American Society of Agricultural Engineers* 107: 221–239.

CIFA (Central Institute of Freshwater Aquaculture) 2009. *Aquaculture Technologies for Farmers*. p. 126.

CRIDA, 2008. Annual Report 2007–08. Central Research Institute for Dryland Agriculture, Hyderabad, India. p. 119.

Desai, A.R. and Adsule, P.G. 2002. Cashew farming systems, an account of their types in Goa region, *In*: Extended Summaries of National Conference on Coastal Agricultural Research, 6–7 April, 2002, ICAR Research Complex for Goa, Old Goa, pp. 122–124.

Dhanda, R.S. 1999. *Department of Forestry and Natural Re-sources*. PAU, Ludhiana (Personal Communications).

Dobermann, A., Witt, C., Dawe, D. 2004. Increasing the Productivity of Intensive Rice Systems through Site-Specific Nutrient Management. Enfield, NH (USA) and Los Baños (Philippines): Science Publishers Inc., and International Rice Research Institute (IRRI); pp. 1–420.

FAO 1993. Guidelines for Land Use Planning. FAO Development Series 1. FAO, Rome, p. 231.

Ganesan, G., Chinnasamy, K.N., Bala Subramanian, A. and Manickasundram, P. 1990. Studies on rice based farming systems with duck cum fish culture in del-taic region of Thanjavur district, Tamil Nadu. *Farm-ing Systems Newsletter* 1(2): 14.

Gangwar, K.S. and Sharma, S.K. 2005. Effect of rice crop establishment methods on the productivity of rice and succeeding chickpea, wheat and mustard. *In*: PDCSR Annual Report 2004–05, Project Directorate for Cropping System Research, Modipuram, India.

Gill, M.S. 2010. Farming system approach: its principles and role towards sustainability. *In*: Extended Summaries, 2010. XIX National Symposium on *"Resource Management Approaches Towards Livelihood Security"*, held during December 2–4, 2010 at Bangalore. pp. 155–159.

Gilley, J.R., Hackbart, C.A., Stetston, L.E. and Feyen, J. 1990. Edited: Book. Management of farm irrigation systems. *American Society of Agricultural Engineers*. pp. 719–745.

GOI. 2008. Department of Animal Husbandry, Dairying and Fisheries, Ministry of Agriculture, GOI, Annual Report 2007–08, p. 111.

Grace, P.R., Jain, M.C. and Harrington L.W. 2002. Environmental concerns in rice-wheat system. *In*: Proceedings of the International Workshop on Developing Action Programme for Farm level Impact in Rice-wheat system of the Indo-Gangetic Plains held during September 25–27, 2000 at New Delhi, India. Rice-Wheat Consortium paper series 14, New Delhi, India, pp. 99–111.

Gupta, R.K. and Abrol, I.P. 1990. Salt–afected soils: Their reclamation and management for crop production. *Advances in Soil Science* 11: 223–288.

Gupta, S.K., Minhas, P.S., Sondhi, S.K., Tyagi, N.K. and Yadav, J.S.P. 2000. Water Resource Management. *In*: (*Ed*.). Yadav, J.S.P. and Singh, G.B. Natural Resource Management for Agricultural Production in India. Lead Papers. International Conference on Managing Natural Resources for Sustainable Agricultural Production in the 21st Century, February, 14–18, 2008, New Delhi, India.

Haider, E., Gangwar, M. and Bhatia, K.S. 2005. Effect of alley width and intra row spacing of Subabul (*Leucaena leucocephala*) on soil loss, growth and yield of Sorghum (*Sorghum bicolor*) under rainfed eco system. *Indian Journal of Agroforestry* 7(2): 15–18.

Hand Book of Agriculture (2009), ICAR, New Delhi

Hegde, B.R., Havanagi, G.V. and Reddy, N.M. 1978. Fertilizer need and water use efficiency of dryland ragi in a cow pea–ragi sequence. *Mysore Journal of Agricultural Sciences* 12: 433–36.

Hegde, B.R., Krishnagouda, K.T. and Parvathappa, H.C. 1988. Organic residue management in red soils under dryland conditions. *In*: Recent Advances in Dryland Agriculture part 2 (*Ed*. Somani, L.L.), Scientific Publishers, Jodhpur, India, pp. 275–85.

Hocking, D. and Rao, G. 1990. Canopy management possibilities for araboreal leucaena in mixed sorghum and livestock small farm production systems in semi-arid India. *Agroforestry Systems* 10: 135–152.

IPCC, 2007. Climate change 2007: The physical science basis. *In*: Solomon, S., Qin, D., Manning, M., Chen, Z., Marquis, M., Averyt, K.B., Tignor, M. and Miller, H.L. (*Eds*.), Contribution of working group I to the Fourth Assessment Report of the Intergovernmental panel on climate change. Cambridge University Press, Cambridge, United Kingdom and New York, NY, USA, p. 996.

Israelson, O.W. 1032. Irrigation Principles and practices, First edition. New York, USA: John Wiley.

Jadhav, S.B., Pawar, S.N., Jadhav, S.B. and Waghmare, A.A. 2006. Evaluation of alley cropping system for soil and water conservation. *Indian Journal of Dryland Agricultural Research and Development* 21(2): 54–156.

James, B.K., Mishra, A., Mohanjty, A., Rajeeb, K., Brahmanand, P.S., Nanda, P., Das, M. and Kiannan, K. 2005. Management of Excess Rainwater in Medium and Lowlands for Sustainable Productivity", Research Bulletin No. 24, Water Technology Center for Eastern Region l (ICAR), Bhubaneswar, India, p. 24.

Jat, M.L., Shrivastava, A., Sharma, S.K., Gupta, R.K., Zaidi, P.H., Rai, H.K., and Srinivasan, G. 2005. Evaluation of maize-wheat cropping system under double no-till practice in Indo-Gangetic basin of India. *In*: 9th Asian Maize Research Workshop, September 4–10, 2005, Beijing, China.

Jat, M.L., Gathala, M., Sharma, S.K., Ladha, J.K., Gupta, R.K., Saharawat, Y.S. and Pathak, H. 2006. Productivity and profitability of rice-wheat system with double No-till practice. International Rice Congress, New Delhi 9–13, October 2006, p. 88.

Jat, M.L., Shrivastava, A., Sharma, S.K., Gupta, R.K., Zaidi, P.H., Rai, H.K. and Srinivasan, G. 2005c. Evaluation of maize-wheat cropping system under double-No-till practice in Indo-gangetic basin of India. *In*: 9th Asian Maize Research Workshop, September 4–10, 2005. Beijing, China.

Jat, M.L. and Pal, S.S. 2000. *In situ* decomposition of rice straw using cellulose-degrading fungi in a typic Ustochrepts of Indo-Gangetic plain under rice-wheat system. (in) Proceedings of National Symposium on Agronomy Challenges and Strategies for the new millennium held during November 15–18, 2000 at Gujarat Agricultural University, Junagadh.

Jat, M.L., Chandana, P., Gupta, R.K., Sharma, S.K. and Gill, M.A. 2006a. Laser land leveling – A precursor technology for resource conservation, *Technical Bulletin*, RWC/CIMMYT, NASC, New Delhi p. 36.

Jat, M.L., Gathala, M., Sharma, S.K., Ladha J.K., Gupta, R.K., Saharawat, Y.S. and Pathak, H. 2006b. Productivity and profitability of rice-wheat system with double No Till practice. International Rice Congress, New Delhi. 9–13 October, 2006, p. 88.

Jat, M.L., Pal, S.S., Shukla, L., Mathur, J.M.S. and Singh, M. 2004. Rice residue management using cellulolytic fungi and its effect on wheat yield and soil health in rice-wheat cropping system. *Indian Journal of Agricultural Sciences*, March.

Jayanthi, C., Rangasamy, A., Mythili, S., Balusamy, M., Chinnusamy, C. and Sankaran, N. 2001. Sustain-able productivity and profitability of integrated farming systems in lowland farms. *In*: Extended summaries. National Symposium on Farming Systems Research in New Millennium, PDCSR, Modipuram. pp. 79–81.

Joshi, P.K., Jha, A.K., Wani, S.P. Joshi, Laxmi and Shiyani, R.L. 2005. Meta analysis to assess the watershed programme and peoples' participation. Comprehensive Assessment Research Report 8, Colombo, Srilanka: Comprehensive Assessment Secretariat, p. 17.

Jacob, J. and Nair, M.A. 1998. An integrated-coconut based mixed farming homestead model for Southern Kerala. *Indian Coconut Journal* 29(8): 4–7.

Jha, K.P., Rao, K.S., Ghosh, A. and Poonam, A. 2001. Rice-based Farming System for Small and Marginal Farmers of Irrigated Areas. *Technical Bulletin*, CRRI, Cuttack.

Kar, G. and Ashwani Kumar. 2007a. Effects of irrigations and straw mulch on water use on tuber yield of potato in eastern India. *Agricultural Water Management*. 94: 109–116.

Kar, G., Ashwani Kumar and Martha M. 2007b. Water use efficiency and crop coefficients of dry season oilseed crops. *Agricultural Water Management*. 87: 73–82.

Kar, G., Sahoo, N., Ashwani Kumar, Das, M. and Roychoudhury, S. 2010. Deep water rice and pond based farming system for enhancing productivity of waterlogged areas. Research Bulletin No. 48. Directorate of Water Management, Bhubaneswar, p. 34.

Kar G., Verma H.N. and Singh, R. 2005. Effects of winter crop and supplemental irrigation on crop yield, water use efficiency and profitability in rainfed rice based cropping system of eastern India. *Agricultural Water Management.* 79(3): 280–292.

Katyal, J.C., Das, S.K., Sharma, K.L. and Saharan, Neelam 1991. Effective nutrient management in improving productivity of dryland crops. Proceedings of FAI Seminar on Optimising Fertilizer Production, Distribution and Usage, S VI/1-12. Fertilizer Association of India, New Delhi.

Katyal, J.C., Reddy, M.N. and Sharma, K.L. 1999b. Current status of crop responses to fertilizers in rainfed areas – an experience from All India Co-ordinated Research Project for Dryland Agriculture. *Fertilizer News* 44(4): 97–112.

Katyal, J.C., Sharma, K.L. and Srinivas, K. 1997. Nutrient management practices in dryland agriculture and cropping systems. *In*: Nutrient Management Practices in Crops and Cropping Systems (*Eds.* Ghonsikar, C.P. and Shinde, V.S.), Scientific Publishers, Jodhpur, India, pp. 191–211.

Katyal, J.C., Sharma, K.L., Narayana Reddy, M. and Neelaveni, K. 1999a. Integrated nutrient management strategies for dryland farming systems. *In*: Fifty Years of Dryland Agricultural Research in India (*Eds.* Singh, H.P., Ramakrishna, Y.S., Sharma, K.L. and Venkateswarlu, B.) CRIDA, Hyderabad. pp. 357–368.

KAU (Kerala Agricultural University) 2007. Package of Practices Recommendations: 'Crops' 13th edn. KAU, Directorate of Extension, Thrissur p. 334.

Korikanthimath, V.S. and Manjunath, B.L. 2009. Integrated farming systems for sustainability in agricultural production. *Indian Journal of Agronomy* 54(2): 140–148.

Kaur, B., Gupta, S.R. and Singh, Gurbachan 2001. Carbon storage and nitrogen cycling in silvipastoral system on sodic soils in north-western India. *Agroforestry Systems* 54: 21–29.

Kelley, W.P. 1951. Alkali soils, their formation, properties and reclamation. Reinhold Publishing Corporation, New York, U.S.A.

Korwar, G.R. 1992. Alternate land use systems. *In*: Somani, L.L. (*Ed.*) Dryland Agriculture in India- State of Art of Research in India. Scientific Publishers, Jodhpur, India. pp. 143–168.

Krishnappa, A.M., Kumar, Arun, Murukannappa, Y.S. and Hegde, B.R. 1999. Improved *in situ* moisture conservation practices for stabilized crop yield in drylands. *In*: (*Eds.*), Singh, H.P., Ramakrishna, Y.S., Sharma, K.L. and Venkateswarlu, B. Fifty years of Dryland Agricultural Research in India. Central Research Institute for Dryland Agriculture (CRIDA), Indian Council of Agricultural Research (ICAR), Hyderabad, pp. 291–300.

Kumar, A. and Abrol, I.P. 1984. Studies on the reclaiming effect of Karnal-grass grown in a highly sodic soil. *Indian Journal of Agricultural Sciences* 54: 189–193.

Lal R. 2004. Soil carbon sequestration on global climate change and food security. *Science*, 304: 1623–1627.

Lal, R. 1997. Long-term tillage and maize monoculture effects on a tropical Alfisol in Western Nigeria. I. Crop yield and soil physical properties. *Soil and Tillage Research*, 42(3): 145–160.

Lal, R. 1989. Agroforestry systems and soil surface management of a tropical alfisol, I: soil moisture and crop yields. *Agroforestry systems* 8: 7–29.

Maji, A.K. 2007. Assessment of degraded and wastelands of India. *Journal of Indian Society of Soil Science*, 55(4): 427–435.

Malik, R.K., Gupta, R.K., Singh, C.M., Yadav, A., Brar, S.S., Thakur, T.C., Singh, S.S., Khan, A.R., Singh, R. and Sinha, R.K. (*Eds.*) 2005. Accelerating the adoption of resource conservation technologies in rice-wheat systems of the IGP. Project Workshop Proceedings, June 1–2, 2005, CCSHAU, Hisar.

Mandal, U.K., Gurcharan Singh, Victor, U.S. and Sharma, K.L. 2003. Green manuring its effect on soil properties and crop growth under rice-wheat cropping system. *European Journal of Agronomy*, 19: 225–237.

Minhas, P.S., Sharma, D.R. and Chauhan, C.P.S. 2004. Management of saline and alkali waters for irrigation. In advances of sodic land recommendation, p. 121–62. International Conference on sustainable management of sodic lands held at Lucknow, Feb. 9–14, 2004.

Mishra, A.K. 2002. Integration of livestock in land use diversification. *In*: Summer School on "Land Use Diversification in rainfed Agro-Ecosystem during 15 April – 5 May 2002, Central Research Institute for Dryland Agriculture, CRIDA.

Mishra, P.K. and Rao, K.V. 1999. Efficient management of surface and groundwater resources in drylands. *In*: (*Eds.*), Singh, H.P., Ramakrishna, Y.S., Sharma, K.L. and Venkateswarlu, B. Fifty Years of Dryland Agricultural Research in India. Central Research Institute for Dryland Agriculture (CRIDA), Indian Council of Agricultural Research (ICAR), Hyderabad. pp. 315–328.

Mishra, A., Sharma, S.D. and Khan, G.H. 2002. Rehabilitation of degraded sodic lands during a decade of Dalbergia sissoo plantation in Sultanpur district of Uttar Pradesh, India. *Land Degradation and Development* 13: 375–386.

Motsara, M.R. 2002. Available nitrogen, phosphorus and potassium status of Indian soils as depicted by soil fertilizer maps. *Fertilizer News*, 47(8): 15–21.

Nanda, S.S. and Garnayak, L.M. 2010. Integrated farming systems for livelihood security in coastal ecosystems. *In*: Extended Summaries, 2010. XIX National Symposium on "*Resource Management Approaches Towards Livelihood Security*", held during December 2–4, 2010 at Bangalore. pp. 160–167.

Nair, P.K.R. 1979. Intensive Multiple Cropping with Coconuts in India. Berlin and Hamburg: Verlag Paul Parrey. p. 135.

Nair, P.K.R. 1985. Classification of agroforestry systems. *Agroforestry Systems* 3: 97–128.

Narain, P., Singh, R.K., Sindhwal, N.S. and Joshi, P. 1998. Agroforestry for soil and water conservation in the western Himalayan valley region of India. *Agroforestry Systems* 39: 191–203.

NBSSLUP, 2004. Soil resource management reports. National Bureau of Soil Survey and Land Use Planning, Nagpur.

NPCC Consolidated Annual Report 2009. "Climate Change Studies conducted at Narendra Dev University of Agriculture & Technology, Kumarganj, Faizabad". pp. 302–312 available at http://www.crida.ernet.in/Climate%20change/network.htm

NRCAF, 2007. NRCAF Perspective Plan – Vision 2025. National Research Centre for Agroforestrry, Jhansi, p.46

Ong, C.K. 1991. Interactions of light, water and nutrients in agroforestry systems. *In Biophysical Research for Asian Agroforestry*. Winrock International (*Eds.* Avery, M.E., Cannell, M.G.R. and Ong, C.K.), New Delhi, pp. 107–124.

Osman, M., Venkateswarlu, B. and Singh, R.P. 1989. Agroforestry research in the tropics of India- A Review. *In*: Singh, R.P., Ahlawat, I.P.S. and Ganga Saran (*Eds.*) National Symposium on Agroforestry Systems in India. Indian Society of Agronomy, New Delhi. pp. 18–35.

Patil, 1994. Annual Report of All India Coordinated Research Project on Salt Affected Soil and Use of Saline Water, Central Soil Salinity Research Institute, Karnal.

PDCSR, 2002. Project Directorate for cropping systems Re-search, Modipuram, Uttar Pradesh (Personal Communication).

Postel, S. 1992. Last Oasis. Worldwatch Institute, Washington DC.

Prasad Rao, G.S.L.H.V., Rao, G.G.S.N. and Rao, V.U.M. (*Eds.*) 2010. Climate Change and Agriculture over India, PHI Learning Pvt. Ltd., New Delhi, p. 328.

Prasad, R.N. 1993. Farming system approach -a solution for sustainable agriculture for hills. *Indian Journal of Hill Farming* 6(1): 19–27.

Prihar, S.S., Gajri, P.R. and Narang, R.S. 1974. Scheduling irrigation to wheat using open pan evaporation. *Indian Journal of Agricultural Sciences* 44: 567–571.

Qadir, M., Qureshi, R.H. and Ahmad, N. 2002. Amelioration of calcareous saline-sodic soils through phytoremediation and chemical strategies. *Soil Use and Management* 18: 381–385.

Lal, R. 2004. Soil carbon sequestration in India. *Climatic Change* 65: 277–296.

Ram, N. 1998. Effect of continuous fertilizer use of soil fertility and productivity of a Mollisol. *In*: Long-term Soil Fertility Management through Integrated Plant Nutrient Supply (Swarup, A., Reddy, D. and Prasad, R.N., *Eds.*), pp. 229–237. All India Research Project on Long-Term Fertilizer Experiments, Indian Institute of Soil Science, Bhopal, India.

Rama Rao W.Y., Tiwari, S.P. and Singh, P. 2005. Crop Livestock Integrated farming system for augmenting socio-economic status of smallholder tribal farmers of

Chhattisgarh in Central India. *Livestock Research for rural Development* 17(8):+ 1–4.

Ramakrishna, Y.S. 1997. Climate features of Indian Arid Zone. *In*: (*Eds*.) Singh, Surendra and Kar, A.). Desertification control in the Arid ecosystem of India for Sustainable Development. Agro-Botanical Publishers, Bikaner. pp. 27–35.

Ramakrishna, Y.S., Rao, G.G.S.N., Rao, V.U.M., Rao, A.V.S. M and Rao, K.V. 2007. Water Management – Water use in rainfed regions of India. *In*: (*Eds*.). Sivakumar, M.V.K. and Motha, R.P.. Managing Weather and Climate Risks in Agriculture, Springer Berlin Hedelberg, New York. pp. 245–264.

Ramakrishna, Y.S., Vittal, K.P.R. and Sharma, K.L. 2005. Conservation agriculture in rainfed semi-arid tropics-some past experiences, lessons learnt and future scope. *In:* Conservation Agriculture-Status and Prospects (*Eds*., Abrol, I.P., Gupta, R.K. and Malik, R.K.), Centre for Advancement of Sustainable Agriculture (CASA), PUSA campus, New Delhi. pp. 199–209.

Rao, D.L.N. and Patra, A.K. 2009. Soil Microbial Diversity and Sustainable Agriculture. *Journal of the Indian Society of Soil Science*, 57(4): 513–530.

Rao, A.C.S. and Das, S.K. 1982. Soil Fertility Management and Fertilizer Use in Dryland Crops. *In*: A decade of Dryland Agricultural Research in India (1971–80). AICRPDA, Hyderabad. pp. 120–39.

Rao, G.G.S.N., Rao, A.V.M., Rao, V.U.M., Vanaja, M., Srinivasarao, M., Rao, K.V., Desai, S. and Srinivasarao, Ch. 2009. Impact, Adaptation and Vulnerability of Climate Change on Rainfed Agriculture. *Indian Journal of Dryland Agricultural Research and Development*, 24: 10–20.

Rao, K.V. and Osman, M. 1994. Silviopastoral Systems for Drylands. Proc. Of International Conference on Sustainable Development of Degraded Lands through agroforestry in Asia and the Pacific, New Delhi, 25–30 Nov., 1994. *In*: Agroforestry systems for Degraded lands (*Eds*. Singh, Panjab, Pathak, P.S. and Roy, M.M.), Oxford and IBH Pub. Co., New Delhi, pp. 755–760.

Rao, M.R., Ong, C.K., Pathak, P. and Sharma, M.M. 1991. Productivity of annual cropping and agroforestry systems on a shallow Alfisol in semi-arid India. *Agroforestry Systems* 15: 51–63.

Rao, U.M.B. and Singh, S.P. 1993. Crop responses to organic sources of nutrients in dryland conditions. *In*: Recent Advances in Dryland Agriculture (*Ed*.) L.L. Somani. Scientific Publishers, Jodhpur, pp. 287–304.

Ravindra Chary, G., Vittal, K.P.R., Ramakrishna, Y.S., Sankar, G.R.M., Arunachalam, M., Srijaya, T. and Udaya Bhanu, 2005. Delineation of soil conservation units (SCUs), soil quality units (SQUs) and Land management units (LMUs) for land resource appraisal and management in rainfed agro ecosystem of India – A conceptual approach. Lead paper. Proceedings of National Seminar on Land Resources appraisal and management for food security, organized by Ind. Soc. of Soil Survey and Land use Planning, NBSSLUP, Nagpur, 10–11, April 2005.

Reddy, D.V.S. and Biddappa, C.C. 2000. Coconut based cropping/farming systems in India. *Journal of Plantation Crops* 28: 1–18.

Reddy, G.H.S. and Reddy, T.Y. 2009. Efficient use of irrigation water. Kalyani Publishers. New Delhi.

Reddy, M.D. 2005. Predominant farming systems and alternatives in Andhra Pradesh. *In*: Singh, A.K., Gangwar, B. and Sharma, S.K. (*Eds*.). Alternative Farming Systems. Farming Systems Research and Development Association, Modipuram. pp. 217–227.

Robbins, C.W. 1986a. Sodic calcareous soil reclamation as affected by different amendments and crops. *Agronomy Journal* 78: 916–920.

Rupakumar, K., Kumar, K. Krishna and Pant, G.B. 1994. Diurnal asymmetry of surface temperature trends over India. *Geophysical Research Letters*., 21: 677–680.

RWC-CIMMYT. 2005. Agenda notes 13th Regional Technical Coordination Committee Meeting. February 6–9, 2005. Dhaka, Bangladesh (RWC-CIMMYT, New Delhi, India)

RWC-CIMMYT. 2006. Research highlights 2005. Rice-wheat consortium for Indo-Gangetic plains, New Delhi, India.

Samra, J.S., Sahoo, N., Roy Chaudhary, S., Mohanty Rajib, K., Jena, S.K. and Verma, H.N. 2003. Sustainable integrated farming system for waterlogged areas of Eastern India. *Research Bulletin* No. 14, WTCER, Bhubaneswar, Orissa, p. 24.

Sehgal, J. and Abrol 1994. Soil degradation in India: Status and impact, New Delhi: Oxford and IBH, p. 80.

Sehgal, J., D.K., Mandal, C. and Vadivelu, S. 1990. Agroecological regions in India, Publication 24, 130 pp. NBSSLUP, Nagpur.

Shanmugasundaram, V.S. and Balusamy, M. 1993. Rice-fish-azolla - An integrated farming system in low-lying wetlands. *Farming Systems* 9: 105–107.

Shanmugasundram, V.S., Balusamy, M. and Rangasamy, A. 1995. Integrated farming system research in Tamil Nadu. *Journal of Farming Systems Research and Development* 1(1 & 2): 1–9.

Sharma, K.L. and Srinivas, K. 1997. Final report of the Indo-US Project on Enhancing Fertilizer Use Efficiency in Conjunction with Residue Management in Dryland Crops and Cropping Systems. CRIDA, Hyderabad. p. 137.

Sharma, K.L., Kusuma Grace, J., Srinivas, K., Ramakrishna, Y.S., Korwar, G.R., Maruthi Sankar, G., Mandal, Uttam Kumar, Ramesh, V., Hima Bindu, V., Madhavi, M. and Pravin, N. Gajbhiye. Influence of Tillage and Nutrient Sources on Yield Sustainability and Soil Quality under Sorghum-Mung Bean System in Rainfed Semi-Arid Tropics. *Communications in Soil Science and Plant Analysis*.

Sharma, K.L., Vittal, K.P.R., Srinivas, K., Venkateswarlu, B. and Neelaveni, K. 1999. Prospects of organic farming in dryland agriculture. *In*: Fifty Years of Dryland

Agricultural Research in India (*Eds*. Singh, H.P., Ramakrishna, Y.S., Sharma, K.L. and Venkateswarlu, B) CRIDA, Hyderabad. pp. 369–78.

Sharma, K.L., Kusuma Grace, J., Mandal, Uttam Kumar, Pravin, N. Gajbhiye, Srinivas, K., Korwar, G.R., Ramesh, V., Ramachandran, Kausalya and Yadav, S.K. 2008. Evaluation of long-term soil management practices through key indicators and soil quality indices using principal component analysis and linear scoring technique in rainfed Alfisols. *Australian Journal of Soil Research*. 46: 368–377.

Sharma, K.L. Mandal, U.K., Srinivas, K., Vittal, K., Mandal, P.R., Kusuma, B., Grace, J. and Ramesh, V. 2005. Long term soil management effects on crop yields and soil quality in dryland Alfisols. *Soil and Tillage Research*. 83: 246–259.

Singh, H.P., Sharma. K.L., Venkateswarlu. B., Vishnumurthy, T. and Neelaveni, K. 1998. Prospects of Indian agriculture with special reference to nutrient management under rainfed systems. *In*: Proceedings of a national workshop on Long-term Soil Fertility Management through Integrated Plant Nutrient Supply (*Eds*. Swarup, A., Reddy, D. Damodar and Prasad, R.N.). Indian Institute of Soil Science, Bhopal, India. pp. 34–53.

Singh, R.P. 1982. Dryland Research in Retrospect and focus in the seventies. *In*: A Decade of Dryland Agricultural Research in India (1971-80), AICRPDA, Hyderabad.

Singh, R.P. 2003. Agroforestry for Conservation of Water Resources. *In*: Agroforestry: Potentials and Opportunities (*Eds*. Pathak, P.S. and Ram Newaj), Agrobios India, pp. 1–6.

Singh, R.P. and Osman, M. 1995. Alternate land use systems for Drylands. *In*: Sustainable Development of Dryland Agriculture in India, Singh, R.P. (*Ed*.) Scientific Publishers, Jodhpur, pp. 375–389.

Singh, B., Sharma, P.K. and Gupta, R.K. 2005. Impact of rice-wheat cropping system on soil and water quality. Proceedings of 'International Conference on Soil, Water and Environmental Quality-Issues and Strategies'. Indian Society of Soil Science, New Delhi.

Singh, Gurbachan, Singh, Harinder, Dagar, J.C., Singh, N.T. and Sharma, V.P. 1997. An evaluation of agri-culture, forestry and agroforestry practices in mod-erately alkali soil in northwestern India. *Agroforestry Systems* 37: 279–295.

Singh, Gurbachan 2005. Integrated farming systems for sustainable soil and water quality. *In*: Proceedings of International conference on soil, water and environmental quality – Issues and strategies. Indian Society of Soil Science, New Delhi, pp. 158–172.

Singh, R.P. and Das, S.K. 1995. Agronomic aspects of plant nutrient management in rain dependent food crop production systems in India. *In*: *Sustainable Development of Dryland Agriculture in India* (*Ed*. Singh, R.P.), Scientific Publishers, Jodhpur. pp. 117–38.

Singh, R.P., Vandenbeldt, R.J., Hocking, D. and Korwar, G.R. 1989. Alley farming in the semi-arid regions of India. *In*: Alley farming in the humid and sub-humid tropics (*Eds*. Kang, B.T. and Reynold, L.), pp. 100–122.

Singh, N.P. and Manjunath, B.L. 2010. Resource management strategies through integrated farming systems for sustainable agricultural production in West coast region. *In*: Extended Summaries, 2010. XIX National Symposium on *"Resource Management Approaches Towards Livelihood Security"*, held during December 2–4, 2010 at Bangalore. pp. 168–170.

Singh, B., Singh, Y. and Sekhon, G.S. 1995. Fertilizer N use efficiency and nitrate pollution of ground water in developing countries. *Journal of Concomitant Hydrology* 20: 167–184.

Singh, J.P. 1996. Studies on agronomical aspects of sequential and simultaneous cropping of sugarcane and wheat in sugarcane-wheat cropping system. p. 18 in Annual Report 1996–97, Modipuram, India: Project Directorate for Cropping Systems Research.

Sinhababu, D.P. and Das, P.C. 2004. Rice-fish Farming System for Rainfed Lowlands Areas. *Technical Bulletin*, CRRI, Cuttack.

Sivnapappan, R.K. 1994. Prospects of micro-irrigation in India. *Irrigation and Drainage System*, (8): 49–58.

Srinivasarao, Ch., Vittal, K.P.R., Chary, G.R., Gajbhiye, P.N. and Venkateswarlu, B. 2006. Characterization of available major nutrients in dominant soils of rainfed crop production systems of India. *Indian Journal of Dryland Agricultural Research and Development* 21: 5105–113.

Srinivasarao, Ch. Ravindra Chary, G., Venkateswarlu, B., Vittal, KPR., Prasad, J.V.N.S., Sumanta Kundu, Singh, S.R., Gajanan, G.N., Sharma, RA., Deshpande, AN., Patel, J.J. and Balaguravaiah, G. 2009a. Carbon sequestration strategies under rainfed production systems of India. Central Research Institute for Dryland Agriculture, Hyderabad (ICAR). p. 102.

Srinivasarao, Ch., and Vittal, K.P.R. 2007. Emerging nutrient deficiencies in different soil types under rainfed production systems of India. *Indian Journal of Fertilizers*, 3: 37–46.

Srinivasarao, Ch., Prasad, J.V.N.S., Vittal, K.P.R., Venkateshwarlu, B. and Sharma, K.L. 2003. Role of optimum nutrition in drought management. *Fertilizer News*. 48(12): 105–112.

Srinivasarao, Ch., Rao, K.V., Chary, G.R., Vittal, K.P.R., Sahrawat, K.L. and Kundu, S. 2009b. Water retention characteristics of various soil types under diverse rainfed production systems of India. *Indian Journal of Dryland Agricultural Research and Development* 24: 1–7.

Srinivasarao, Ch., Venkateswarlu, B., Dixit, S., Kundu, S. and Gayatridevi, K. 2011a. Livelihood impacts of soil health improve-ment in backward and tribal districts of Andhra Pradesh. Central Research Institute for Dryland Agriculture. Hyderabad. p. 121.

Srinivasarao, Ch., Venkateswarlu, B., Sharma, K.L., Ramachandrappa, B.K., Patel, J.J. and Deshpande, A.N. 2011b. Analyzing nitrogen use efficiency in long term manurial experiments under rainfed production systems of India. *Indian Journal of Fertilizers* p. 7.

Srinivasarao, Ch., Vittal, K.P.R., Gajbhiye, P.N., Kundu, Sumanta and Sharma, K.L. 2008. Distribution of micronutrients in soils in rainfed production systems of India. *Indian Journal of Dryland Agricultural Research and Development*, 23: 29–35.

Srinivasarao, Ch., Vittal, K.P.R., Kundu, S., Gajbhiye, P.N. and Vijayasankarbabu, M. 2010b. Twenty Years of Continuous Cropping, Fertilization and Organic Manure Application Effects on Forms of Potassium and Kinetics of Nonexchangeable K Release in the Profile of an Alfisol Under Arid Conditions. *Communications in Soil Science and Plant Analysis*, 41(6): 783–796.

Srinivasarao, Ch., Vittal, K.P.R., Venkateswarlu, B., Wani, S.P., Sahrawat, K.L., Marimuthu, S. and Kundu, S. 2009c. Carbon stocks in different soils under diverse rainfed production systems in tropical India. *Communications in Soil Science and Plant Analysis*, 40(15): 2338–2356.

Srinivasarao, Ch., Wani, S.P., Sahrawat, K.L. and Rajasekhara Rao, B.K. 2009d. Nutrient management strategies in participatory watersheds in semi arid tropical India. *Indian Journal of Fertilizers* 5(12): 113–128.

Srinivasa Rao, Ch., Venkateswarlu, B., Srinivas, K., Sumanta Kundu. and Anil Kumar Singh. 2011d. Soil Carbon Sequestration for Climate Change Mitigation and Food Security. Central Research Institute for Dryland Agriculture, Hyderabad, India, p. 322.

Srinivasarao, Ch., Venkateswarlu, B., Lal, R., Singh, A.K., Sumanta Kundu, Vittal, K.P.R., Patel, J.J. and Patel, M.M. 2011. Long-term manuring and fertilizer effects on depletion of soil organic carbon stocks under pearl millet-cluster bean-castor rotation in western India. Land Degradation and Development. DOI: 10.1002/ldr.1158.

Srinivasarao, Ch., Venkateswarlu, B., Lal, R., Singh, A.K., Vittal, K.P.R., Sumanta Kundu, Singh, S.R. and Singh, S.P. 2012a. Long-term effects of soil fertility management on carbon sequestration in a rice-lentil cropping system of the Indo-Gangetic plains. Soil Science Society of America Journal. 76(1), doi: 10.2136/sssaj2011.0184.

Srinivasarao, Ch., Venkateswarlu, B., Singh, A.K., Vittal, K.P.R., Sumanta Kundu, Gajanan, G.N., Ramachandrappa, B. and Chary, G.R. 2012b. Critical carbon inputs to maintain soil organic carbon stocks under long term finger millet (*Eleusine coracana* (L.) Gaertn) cropping on Alfisols in semi arid tropical India. Journal of Plant Nutrition and Soil Science (In press).

Subba Rao, N.S. 2007. Biofertilizers in Agriculture. Publisher: Routledge. p. 200.

Subba Reddy, G. and Ramakrishna, Y.S. 2005. Farming Systems Modules in Rainfed Agriculture. *In*: Singh, A.K., Gangwar, B. and Sharma, S.K. (*Eds.*). Alternative Farming Systems. Farming Systems Research & Development Association Modipuram. pp. 250–110.

Swarup, A. 1998. Emerging soil fertility management issues for sustainable crop productivity in irrigated systems. *In*: Long-term Soil Fertility Management through Integrated Plant Nutrient Supply (*Eds.* Swarup, A., Reddy, D. Damodar and Prasad, R.N.), All India Coordinated Research Project on Long-term Fertilizer Experiments, Indian institute of Soil science, Bhopal , India. pp. 54–68.

Tandon, H.L.S. 1987. Phosphorus in dryland agriculture. *In*: *Phosphorus Research and Agricultural Production in India* (*Ed.* Tandon, H.L.S.), Fertilizer Development and Consultation Organisation, New Delhi, pp. 73–84.

Tandon, H.L.S. 1993. *Fertilizer Management in Rainfed Dryland Agriculture – a Guidebook.* Fertilizer Development and Consultation Organisation, New Delhi. pp. 104–8.

TAR-IVLP, 2003. Annual report of technology assessment and refinement through institute village linkage program (TAR-IVLP) under rainfed agro-ecosystem. Central Research Institute for Dryland Agriculture, Hyderabad, p. 94.

TAR-IVLP, 2004. Annual report of technology assessment and refinement through institute village linkage program (TAR-IVLP) under rainfed agro-ecosystem. Central Research Institute for Dryland Agriculture, Hyderabad, p. 90.

TAR-IVLP, 2005. Completion report of technology assessment and refinement through institute village linkage program (TAR-IVLP) under rainfed agro-ecosystem. Central Research Institute for Dryland Agriculture, Hyderabad.

Timsina, J. and Connor, D.J. 2001. Productivity and management of rice-wheat cropping systems: Issues and challenges. *Field Crops Research* 69: 93–132.

Varughese, K. and Mathew, T. 2009. Integrated farming systems for sustainability in coastal ecosystem. *Indian Journal of Agronomy* 54(2): 120–127.

Velayutham, M., Mandal, D.K., Mandal, Champa and Sehgal, J. 1999. Agroecological subregions of India for planning and development. NBSS Publ. 35, 372. NBSSLUP, Nagpur, India.

Velayutham, M., and Bhattacharyya, T. 2000. Soil Resource Management. *In*: Natural Resource Management for Agricultural Production in India (*Eds.*) Yadav, J.S.P. and Singh, G.B. International Conference on Managing Natural Resources for Sustainable Agricultural Production in the 21st Century. Feb. 14–18, 2000, New Delhi, India: 3–135.

Venkateswarlu, B. 2010. Climate Change: Adaptation and Mitigation Strategies in Rainfed Agriculture. The 21st Dr. S.P. Raychaudhuri Memorial Lecture, *Journal of Indian Society of Soil Science*, 58.

Venkateswarlu, B. 2007. Nitrogen fixation in arid and. semi-arid agriculture: Opportunities and constraints. (*In*: Abrol, Y.P., Raghuram, N. and Sachdev, M.S. (*Eds.*)), Agricultural Nitrogen Use and Its Environmental Implications. I.K. International Publishing House, Pvt., Ltd. New Delhi. pp. 415–436.

Venkateswarlu, B., Srinivasarao, Ch., Ramesh, G., Venkateswarlu, S. and Katyal, J.C. 2007. Effects of long term legume cover crop incorporation on soil organic carbon, microbial biomass, nutrient build up and grain yields of sorghum/ sunflower under rainfed conditions. *Soil Use and Management* 23: 100–107.

Venkateswarlu, B. and Wani, S.P. 1999. *Bio-fertilizers: An Important Component of Integrated Plant Nutrient Supply in Drylands. In*: Fifty Years of Dryland Agricultural Research in India (*Eds*. Singh, H.P., Ramakrishna, Y.S., Sharma, K.L. and Venkateswarlu, B.). Central Research Institute for Dryland Agriculture. Hyderabad, pp. 379–394.

Venkateswarlu, J. 1986. *In*: 15 years of Dryland Agricultural Research, Souvenir. CRIDA, Hyderabad, pp. 42–58.

Venkateswarlu, J. 1987. Efficient resource management systems for drylands of India. *Advances in Soil Science*, 7: 165–221.

Vineela, C., Wani, S.P., Srinivasarao, Ch., Padmaja, B. and Vittal, K.P.R. 2008. Microbial properties of soils as affected by cropping and nutrient management practices in several long-term mnaurial experiments in the semi arid tropics of India. *Applied Soil Ecology*, 40: 165–173.

Vittal, K.P.R, Rao, K.V., Sharma, K.L, Victor, U.S, Ravindra Chary, G., Maruthi Sankar, G.R., Samra, J.S. and Singh, Gurbachan 2005. Agricultural Drought Zonation, cropping practices and amelioration paths for rainfed production systems. *Indian Journal of Dryland Agricultural Research & Development.* 20(2): 155–202.

Vittal, K.P.R., Ravindra Chary, G., Rama Rao, C.A. and Sankar, G.R.M. 2006. Crop diversification in rainfed regions of India. *In*: (*Eds*.) Sharma, K.D. and Soni, B. Land use diversification for sustainable agriculture. Atlantic publishers and distributors, New Delhi pp. 24–70.

Vittal, K.P.R., Singh, H.P., Rao, K.V., Sharma, K.L., Victor, U.S., Chary, G.R., Sankar, G.R.M., Samra, J.S. and Singh, Gurubachan (*Eds*.) 2003. Guidelines on drought coping plans for rainfed production systems. All India Coordinated Research Project for Dryland Agriculture, Central Research Institute for Dryland Agriculture, Indian Council of Agricultural Research, Hyderabad, p. 39.

Yadav, J.S.P. 1984. Presidential Address, 49th Annual General Meeting, Indian Society of Soil Science. *Journal of Indian Society of Soil Science*. 32(4): 551–555.

Yadav, R.L. 1981. International pigeon pea to conserve fertilizer nitrogen in maize and produce residual effect on sugarcane. *Experimental Agriculture* 17: 311–315.

Yadav, R.L. 1982. Minimizing nitrate-nitrogen leaching by parallel multiple cropping in long duration crops. *Experimental Agriculture* 18: 37–42.

Yadav, R.L., Singh, S.R., Prasad, K., Dwivedi, B.S., Batta, R.K., Singh, A.K., Patil, N.G. and Chaudhary, S.K. 2000. Management of irrigated ecosystem, *In*: (*Eds*.). Yadav, J.S.P. and Singh, G.B. *Natural Resource Management for Agricultural Production in India, Lead papers. International Conference on Managing Natural Resources for Sustainable Agricultural Production in the 21st Century, February, 14-18, 2000, New Delhi*. pp. 775–870.

ANNEXURE 1

Template for Agriculture Contingency Plan State

Andhra Pradesh; District: Krishna

1.0 District Agriculture profile

1.1 Agro-Climatic/Ecological Zone

1.2 Rainfall	*Normal RF (mm)*	*Normal Rainy days (no)*	*Normal Onset (specify week and month)*	*Normal Cessation (specify week and month)*

1.3 **Land use pattern of the district** (latest statistics) (Give year of data)	Geographical Area	Forest area	Land under non-agricultural use	Permanent pastures	Cultivable wasteland	Land under Misc.tree crops and groves	Barren and unculti-vable land	Current fallows	Other fallows

1. 4 Major soils (common names like shallow red soils etc.,)	Area ('000 ha)	Percent (%) of total
1.5 Agricultural land use	Area ('000 ha)	Cropping intensity %
1.6 Irrigation		

Area under major field crops & horticulture etc. (2008-09)

1.7 Major Field Crops cultivated					**Area ('000 ha)**			
1.8 Livestock		**Male ('000)**			**Female ('000)**			**Total ('000)**
1.9 Poultry		**No. of farms**			**Total No. of birds ('000)**			
1.10 Fisheries (Data source: Chief Planning Officer)								
1.11 Production and Productivity of major crops (Average of last 5 years: 2004, 05, 06, 07, 08)	**Kharif**		**Rabi**		**Summer**		**Total**	**Crop residue as fodder ('000 tons)**
	Production ('000 t)	Productivity (kg/ha)	Production ('000 t)	Productivity (kg/ha)	Production ('000 t)	Productivity (kg/ha)	Production ('000 t)	Productivity (kg/ha)
1.12 Sowing window for 5 major field crops (start and end of normal sowing period)								
	Crop 1: Paddy		2: Maize		3: Pulses		4: Groundnut	5: Sugarcane
1.13 What is the major contingency the district is prone to? (Tick mark and mention years if known during the last 10 year period)								
	Regular				Sporadic			None
1.14 Include Digital maps of the district for								
	Location map of district within State as Annexure Mean annual rainfall as Annexure Fertility Status as Annexure							

2.0 Strategies for weather related contingencies
2.1 Drought

2.1.1 Rainfed Situation

Early Season Drought		Suggested Contingency measures			
Early season drought (delayed onset) (2, 4, 6 and 8 weeks)	*Major Farming situation*[a]	*Normal Crop/ cropping system*[b]	*Change in crop/ cropping system*[c]	*Agronomic measures*[d]	*Remarks on Implementation*[e]
Mid-season Drought					
Early season drought (Normal onset) Normal onset followed by 15-20 days dry spell after sowing leading to poor germination/crop stand etc. Mid season drought (long dry spell, consecutive 2 weeks rainless (>2.5 mm) period)At vegetative stage reproductive stage					
Terminal Drought					

21.2 Irrigated situation

Condition	Suggested Contingency measures				
	Major Farming situation[a]	*Normal Crop/ cropping system*[b]	*Change in crop/ cropping system*[c]	*Agronomic measures*[d]	*Remarks on Implementation*[e]
Delayed release of water in canals due to low rainfall					
Limited release of water in canals due to low rainfall					
Non release of water in canals under delayed onset of monsoon in catchment					
Lack of inflows into tanks due to insufficient /delayed onset of monsoon Insufficient groundwater recharge due to low rainfall					

2.2 Unusual rains (untimely, unseasonal etc) (for both rainfed and irrigated situations)

Condition - Continuous high rainfall in a short span leading to water logging				
Crop	**Suggested contingency measure**			
	Vegetative stage[k]	*Flowering stage*[l]	*Crop maturity stage*[m]	*Post harvest*[n]

2.3 Floods

Condition	**Transient water logging/ partial inundation**[1] **Suggested contingency measure**[o]			
	Seedling / nursery stage	*Vegetative stage*	*Reproductive stage*	*At harvest*
Condition - Continuous submergence for more than 2 days[2]				

2.4 Extreme events: Heat wave/Cold wave/Frost/ Hailstorm	**Suggested contingency measure**[r]			
	Seedling/nursery stage	*Vegetative stage*	*Reproductive stage*	*At harvest*
Heat wave				
Cold wave				
Frost				
Hailstorm				
Cyclone				

2.5 Contingent strategies for Livestock, Poultry & Fisheries

2.5.1 Livestock

	Suggested contingency measures		
	Before the event[s]	**During the event**	**After the event**
Drought			
Floods			
Cyclone			
Heat wave and cold wave			

2.5.1.1 Poultry

	Suggested contingency measures		Convergence/linkages
	Before the event[a]	**During the event**	**After the event**
Drought			
Floods			
Cyclone			
Heat wave and cold wave			

2.5.3 Fisheries/ Aquaculture

	Suggested contingency measures		
	Before the event[a]	**During the event**	**After the event**
1. Drought			
2. Floods			
3. Cyclone / Tsunami			
4. Heat wave and cold wave			

SUBJECT INDEX

D

E

F

G

H

I

K

L